SMITHSONIAN ANSWER BOOK

SNAKES

SMITHSONIAN ANSWER BOOK

SNAKES

GEORGE R. ZUG AND CARL H. ERNST

PHOTOGRAPHS BY

RICHARD AND PATRICIA BARTLETT

SMITHSONIAN BOOKS
WASHINGTON

Copyeditor: Fran Aitkens
Production Editor: Robert A. Poarch
Designer: Janice Wheeler

Library of Congress Cataloging-in-Publication Data
Zug, George R., 1938–

    Snakes : Smithsonian answer book / George R. Zug and Carl H. Ernst.—2nd ed.

        p.   cm.

    Rev. ed. of: Snakes in question / Carl H. Ernst and George R. Zug ; illustrations by Molly Dwyer Griffin. c1996.

    Includes bibliographical references (p.).

ISBN 1-58834-113-5 (alk. paper) — ISBN 1-58834-114-3 (pbk. : alk. paper)

    1. Snakes—Miscellanea. I. Ernst, Carl H.   II. Ernst, Carl H. Snakes in question.   III. Title.

QL666.O6E48 2004

597.96—dc22   2003059020

British Library Cataloguing-in-Publication Data are available.

Manufactured in China
10 09 08 07 06 05 04   5 4 3 2 1

♾ The paper used in this publication meets the minimum requirements of the American National Standard for Information Sciences—Permanence of Paper for Printed Library Materials ANSI Z39.48-1984.

Frontispiece: **Cape cobra (*Naja nivea*)**

To biologists and naturalists, past and present,

who have gathered the facts about snakes,

and to our wives, Pat and Evelyn,

for years of support.

CONTENTS

.1.
SNAKE FACTS

.2.

FOLKTALES

.3.
GIANT SNAKES: BIG AND BIGGEST

.4.
SNAKEBITE

.5.

SNAKES AND US   129

PREFACE

Snakes fascinate, thrill, and even repel us. They fascinate us because they are limb-less but can do many things we would not expect them to do. They thrill us because some are gigantic, others are venomous, and all represent the mystery of the un-known. But why do they repel us? The reason for our repulsion is unclear. Human beings may be innately wary of serpentlike creatures, but repulsion is probably a learned response. It probably arises from folktales and exaggeration, and perhaps from the association of the metaphorical serpent in the Garden of Eden with actual snakes. Childhood associations of snakes with danger and evil become part of a mental outlook, more subconscious than conscious, that can be changed only with effort and access to accurate information. Thus our first goal in writing this book is to present accurate information. We believe that understanding snakes makes them less frightening and threatening. And the more they are appreciated as normal members of natural communities, the less likely they are to be killed out of fear.

Snakes in Question sets out to answer the most frequently asked questions about snakes. Organized in question-and-answer format like its predecessor *Sharks in Question*, it also offers insight into the basic biology of snakes. Snake biology is full of strange and unexpected facts, few of which are widely known. Although the book is aimed at North American readers, its scope is worldwide. We assume that readers have heightened curiosity about snakes, perhaps stimulated by encounters with snakes in zoos or in the wild, or by television nature programs. Like such pro-grams, this book also presents strange and spectacular facts, but we have tried to do so in the context of the basics of snake biology and diversity. "Snake Facts" presents those basics. In "Folktales" we examine some common misunderstandings about snakes and attempt to trace how such misunderstandings arose from incomplete

Mexican speckled racer (*Drymobius margaritiferus*)

observation and how observational errors are compounded by misinterpretation. "Giant Snakes: Big and Biggest" focuses on awesomely gigantic snakes, and "Snake-bite" looks at the danger, often overstated but nonetheless real, of a snakebite. In "Snakes and Us" we examine the relationship between snakes and humans. Because we expect most readers to pursue questions that interest them rather than reading sequentially, some information appears in several sections.

Much has been written about snakes. We have compiled two types of bibliographies to guide the readers who want to know more. The general bibliography is a list of popular and broader-scope books about snakes and their biology. The subject bibliography gives lists of recommended further readings on each of the major topics covered in this book. It is organized like the text; publications are listed according to the particular question about snakes that they address.

ACKNOWLEDGMENTS

FIRST EDITION

Peter Cannell first suggested that we write about snakes for the second Smithsonian Answer Book. The question-answerers (Ron Crombie, Steve Gotte, Bob Reynolds, and Addison Wynn, all with the National Museum of Natural History, Division of Amphibians and Reptiles) gave us encouragement and ideas, and then reviewed the entire first draft of the manuscript. Joy Gold, Duane Hope, and Van Wallach also read and commented on the entire manuscript. Other colleagues, including Kraig Adler, Tom Anton, David L. Auth, William S. Brown, Peter Cannell, David Cundall, Herndon G. Dowling, Carl Gans, David L. Hardy, Robert W. Henderson, J. Alan Holman, Fran Irish, Karl V. Kardong, Ernest A. Liner, Sherman A. Minton, Alan Resetar, Jose P. Rosado, Richard A. Seigel, Joseph B. Slowinski, Peter D. Strimple, and Bruce A. Young, have read various sections, ferreted out errors, and offered helpful suggestions. Our wives, Evelyn and Pat, encouraged us throughout and occasionally lent a hand to type portions of the text. We appreciate the generous assistance of all, and thank them for making this a better book. We alone remain responsible for any remaining errors.

SECOND EDITION

With the continual appearance of snake books, often with more glitch than accuracy, Vince Burke recognized the need to update *Snakes in Question* in a more eye-catching format through the addition of color photographs. We were delighted with his suggestion because it gave us an opportunity to do some fine-scale tweaking of the text and to present some of the new ideas concerning snake origins and classification. We remain in debt to the reviewers of and contributors to the first edition and to Ann Goodsell, our former editor, who created readable prose without losing scientific accuracy. Our current editor, Fran Aitkens, also has polished our prose; Nicole Sloane of Smithsonian Books has coordinated image selection and other must-do aspects for publication of this edition. Thanks to both! We also wish to thank Kerry Hanskecht for providing snake diving data and Jason Head for the latest information on fossil snakes and assistance with color preparation of the line art.

Most of the photographs are by Dick and Pat Bartlett. When they were unable to provide the desired images, other colleagues graciously helped. Their images are identified by their names in parentheses at the end of the image legends.

INTRODUCTION

There are more than 2,800 species of snakes in the world, including some as yet unknown to science. Many are rare, and a few have been seen only once. Others are familiar to us, whether we live in the city or the country. Snakes are found from the Arctic Circle to the equator, and southward to the southern tips of all the continents except South America; they are totally absent only from Antarctica. Some live in trees, others underground; some live in deserts and others in oceans. Some are so small as to be mistaken for earthworms, others so large as to worry humans who fear becoming their next meal. Most are nonvenomous, but the bites of a few are deadly without medical care.

The diversity of snakes is part of their fascination. Indeed, some aspects of snake biology are so astonishing as to compare with myths and tall tales. There are snakes with light sensors on their tails, snakes with infrared sensors, snakes with hinged teeth, snakes that are immune to the toxic venoms of other snakes. Some eat only fish eggs or bird eggs, some lay eggs and others give birth to more than 50 newborns at a time. This is only a sampling of remarkable facts about snakes.

The diversity of snakes is an evolutionary success story that began more than 130 million years ago. In spite of the human propensity to bludgeon every snake in sight, there are still many snakes today. We limbed beings continue to marvel at these creatures that travel from place to place without limbs and catch and eat prey without arms and hands. The reasons for some of their success will become obvious in the pages that follow.

.1.

SNAKE FACTS

WHAT ARE SNAKES?

Snakes are limbless or near-limbless reptiles. Their relatives include turtles, croco-dilians, birds, and lizards among living reptiles, and dinosaurs, pterodactyls, and plesiosaurs among the extinct ones. Snakes are also vertebrates, and like all other vertebrates (sharks, bony fishes, frogs, and mammals) they have a segmented spinal column of bony vertebrae.

Snakes share with their reptilian relatives a skin covered with horny (kerati-nous) scales, an external nasal gland, and other unique structural traits. Some of these traits cannot be seen without a microscope or dissection, but because the traits have persisted in turtles, crocodilians, lizards, and even birds, and are absent in mammals and amphibians, most biologists regard the snakes and other reptiles (including birds) as a single evolutionary lineage—the Reptilia (see Appendix 1).

HOW DO SNAKES DIFFER FROM LIZARDS?

No one is likely to mistake a turtle or a crocodile for a snake. But even herpetolo-gists have mistaken legless lizards for snakes and vice versa. Some similarities in ap-pearance are the shared genetic heritage of a common ancestor, others result from evolutionary convergence in behavior and anatomy. Because some of the genetic similarities are unique to snakes and lizards, we know that snakes are more closely

Olive forest racer (*Dendrophidion dendrophis*)

1

As in snakes, glass lizards (*Ophisaurus*) are limbless, but unlike snakes, these legless lizards have moveable eyelids. Glass lizards are so named because two-thirds of the lizard is tail that often breaks into several pieces when grasped by a predator.

related to lizards than to any other reptile. Indeed, in a strict evolutionary sense, snakes are limbless lizards, although they are not closely related to any living limbless lizards, such as the glass lizards, slow-worms, or worm lizards. Snakes first appeared millions of years ago (see *When Did the First Snake Appear?*).

At first glance, snakes seem easy to distinguish from lizards, but the great diversity of lizards—more than 4,000 different species—has produced many convergent adaptations, such as loss of limbs and eyelids. The unique traits of snakes are deep anatomical ones, not observable with the naked eye. Snakes' unique traits include a nonfunctional or absent left lung, presence of a tracheal lung, a specialized retina, and a lower jaw whose two sides are separate in the front. In most snakes, the ventral (underside of the body) scales are enlarged and arranged in a single longitudinal series, like a tread on a bulldozer. The eyes are well developed in most snakes, but moveable eyelids have been lost. Instead, each eye is covered by a single transparent scale called the *spectacle*. Snakes have middle and inner ears, but no external ear openings.

Boas and pythons have remnants of their hindlimbs projecting as a horny spur on either side of the opening of the cloaca. The cloacal spur of this female ball python (*Python regius*) is small relative to the spurs of males.

No snake has any remnant of the forelimbs, shoulder girdle, or sternum. The lack of forelimbs and pectoral girdle also characterizes some limbless or reduced-limb lizards and thus is not a reliable trait for distinguishing snakes from lizards. Most snakes also lack any bones, no matter how small, of the hindlimbs and pelvic girdle. However, members of older snake groups (boas, pythons, blindsnakes) do have remnants of the hindlimbs and pelvis. In boas and pythons, the hindlimb remnants are visible externally as the cloacal spurs, one on each side of the body immediately in front of the base of the tail. Each spur contains a tiny cylinder of bone, the femur, articulating to a small bony chip, the pelvis. These limb and girdle remnants have remnants of limb muscles and can be moved. In boas and pythons, males typically have larger spurs than females, and the male uses his spurs to stimulate the female during courtship.

HOW ARE SNAKES BUILT?

Snakes come in all sizes, from pygmies less than 15 centimeters long to giants more than 6 meters long. All share an elongate body with no limbs, except for a small horny spur on each side of the vent in the few species with remnants of hindlimbs. The body contains a string of vertebrae, typically more than 120 in the body and tail and in some species as many as 585 (Oenpelli python, *Morelia oenpelliensis*). Each neck and trunk vertebra bears a pair of ribs, except for the first two vertebrae (the atlas and axis), which respectively attach the vertebral column to the skull and allow rotation of the head. The ribs are slender but sturdy. Each rib pair encircles more than half of the trunk, ending on the underside at the edge of a large ventral scale. Because most snakes (except Typhlopidae) have a pair of ribs corresponding to each ventral scale, the number of ventral scales usually equals the number of neck

Scale covering on the dorsal surface of the head. Many vipers have small head scales (right), Mt. Kenya bushviper (*Atheris desaixi*), and most colubrids have large head scales (left), Boulenger's horn-nosed snake (*Rhynchophis boulengeri*).

and trunk vertebrae. The relationship between ribs and ventral scales is an important feature of snake locomotion (see *How Do Snakes Crawl and Swim?*).

The number of caudal (tail) vertebrae and ribs is variable, depending on the length and function of the tail. Generally, ribs only occur on the first few tail vertebrae.

Snakes wear a sheath of dry keratinous scales. The arrangement, size, and number of scales vary from species to species and by location on the body. Generally speaking, the head has large scales on top, somewhat smaller scales on the side, and smaller scales around the mouth and under the chin. Skin flexibility varies accordingly, with little stretching necessary or possible on the top, but extreme elasticity of the skin around the mouth and throat. Immediately behind the head, the scales become small. The scales of the body usually lie in regular linear rows, diagonal to the long axis of the body. Each row has a fixed number of scales, typically an odd number ranging from 13 to 27. The number is species-specific and can be the same behind the head, at midbody, and in front of the tail, such as 13–13–13 (eastern wormsnakes, *Carphophis amoenus*), or can decrease in a regular pattern, such as 25–23–17 (American ratsnakes, *Elaphe obsoleta*). The number of scales in a row tends to vary with body girth: thin-bodied snakes usually have fewer scales per row than thick-bodied snakes. However, some thick-bodied species have only a few large scales, and the reverse is true for some slender species. Scale number does not change with growth. The number and arrangement of scales is the same at hatching and in adulthood.

On the underside or venter of most snakes, large rectangular scales are arranged in a single row from the throat to the vent. This single row of ventral scales is critical for locomotion; these scales provide the contact edges for pushing the sides and top of the body forward. Some stretch and spread is possible between the ventral scales, but the more numerous rows of scales on the sides of the body permit maximum expansion of the body wall to accommodate large meals in the gut. Of course, not all snakes have large ventral scales. Seasnakes' scales are small as an adaptation to reduce friction in their marine environment and blindsnakes' scales are all nearly equal in size, perhaps as protection from the sharp jaws of their ant and termite prey.

The skin is loosely yet regularly attached to the body musculature. It can stretch forward–backward or up–down on the body to accommodate swallowed prey, yet remain fixed enough to contact a surface for locomotion.

The integument can stretch because each scale overlaps those behind and beside it. A soft skin lies between each set of scales and beneath the overlap. This division

The undersides of most snakes have transversely arranged enlarged scales, the ventral scales. These scales give critical traction for undulatory locomotion. As in this red-bellied watersnake (*Nerodia erythrogaster*), the ventral scales are often brightly colored.

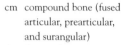

cm compound bone (fused
     articular, prearticular,
     and surangular)
d   dentary bone
ec   ectopterygoid bone
f   frontal bone
m   maxillary bone
n   nasal bone
pa   parietal bone
pl   palatine bone
pm   premaxillary bone
po   postorbital bone
prf prefrontal bone
pt   pterygoid bone
q   quadrate bone
st   supratemporal bone

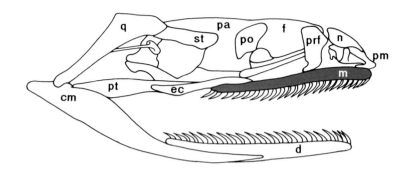

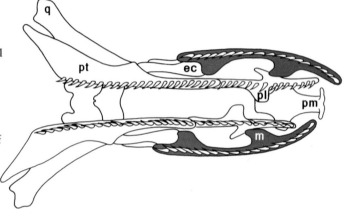

Figure 1.1. Representative snake skull (skull of a brown watersnake, *Natrix taxispilota*). The lateral view shows the major bones in the snake's skull. In the ventral view the lower jaw is removed to show how the two sides of the upper jaw bone move independently. (Adapted from Cundall and Gans 1979)

of the scales allows the integument to stretch over large prey as it passes rearward through the gut. In fat captive snakes and recently fed snakes, the scales are widely separated by the stretching of skin between them.

Muscles are the motile force lying between the skeleton and the skin. Without limbs, body movement provides locomotion through the muscle system's complex linkages with the ribs and vertebrae. Muscles link adjacent vertebrae to ribs, and more distant vertebrae to one another, as well as linking vertebrae and ribs to the ventral scales. These complex linkages give snakes' bodies extraordinary flexibility, a range of locomotive mechanisms, and, in many snakes, the ability to constrict prey.

Snakes' many vertebrae and ribs are linked by a body musculature that functions like a complex set of ropes (muscles) and pulleys (bones) to produce their undulatory locomotion and other movements. Each vertebra typically has muscular links to the two or three vertebrae ahead of and behind it, as well as muscle slips to one or more ribs and the skin. The resulting complex overlap of muscle slips gives the snake its great flexibility and movement control. The muscles of the head, although less complex, form a sophisticated mechanism to move the many parts of the jaw.

Snake skulls (Figure 1.1) consist of three functional units: a braincase (cranium), sensory capsules, and the jaw apparatus. The braincase is a bony box protecting the brain and providing a surface for the attachment of muscles and various jaw bones. The sensory capsules, which house the "nose," eyes, and ears, are strategically attached to the braincase.

The jaw apparatus on each side of the head consists of two or three units in the upper jaw and a lower jaw unit. Each unit can move independently both of adjacent bones on the same side of the head and of its counterpart on the opposite side of the head; it can also move in coordination with them. This independence gives the skull a flexibility unmatched in other vertebrates, and permits snakes to catch and swallow large food items (see *What and How Do Snakes Eat?*). As a group, snakes consume larger prey relative to their body size than most other vertebrate predators. Lacking limbs and cutting teeth, snakes cannot hold or tear apart their prey and so must swallow it whole.

The digestive tract is a long tube, all of whose parts are expandable, that runs from the mouth to the anus. The mouth with its numerous teeth serves to catch, hold, and engulf food items. The teeth also puncture the food item, allowing digestive enzymes to enter into the food. Salivary glands empty saliva (mucus) into the mouth cavity; the saliva coats the food and eases its passage down the esophagus and into the stomach, which is extremely muscular. The food begins its breakdown via stomach acids and enzymes and muscular "crushing." After the major physical and initial chemical breakdown, the food passes from the stomach into a relatively short small intestine for further chemical breakdown and absorption of its nutrients. It then moves into the large intestine or colon for reabsorption of water, and exits through the anus into the cloaca and finally to the outside through the vent— the transverse slit at the beginning of the tail. The cloaca is a shared receptacle into which the digestive, excretory, and reproductive systems empty their products. (For a description of reproductive organs, see *How Do Snakes Reproduce?*)

The excretory system consists of a pair of kidneys (metanephric type). Each kidney is drained by a ureter emptying directly into the cloaca; snakes lack a urinary bladder. Urine or nitrogenous waste is passed as a semiliquid paste of uric acid crystals, hence its similarity to bird excreta. Uric acid excreta is nontoxic and low in water content—an adaptation to conserve water.

Oxygen is delivered to and carbon dioxide removed from body tissue by blood flowing through the circulatory system. The snake heart is three-chambered: two atria collect unoxygenated blood from the body and oxygenated blood from the lungs, and a ventricle pumps blood to the lungs and remainder of the body. The arteries and veins are organized as in other reptiles and mammals, although obviously lacking the major vessels to the limbs. One major difference is a vertebral plexus in those snakes that regularly climb or crawl with raised heads, such as the American whipsnakes (*Masticophis*). This plexus is a network of spinal veins within the vertebral column interconnected with caval and portal veins lying beneath the column. When a snake is horizontal, the caval and portal veins are the major blood channels; however, climbing or crawling with an elevated head causes loss of blood pressure in these vessels, so the plexus ensures adequate blood flow and pressure to the head.

HOW DO SNAKES BREATHE?

Snakes have lungs. All snakes have a right lung, and some also have a left one, although the left lung is often small. The right lung is large, extending rearward to and often beyond the middle of the body. The length of the right lung is variable among species but consistent within a species. For example, it is short in one species and nearly the full length of the body cavity in another species (*Acrochordus, Pelamis*).

In many species with either a normal-sized or a reduced left lung, a new lung-like structure has appeared. The new structure develops from the wall of the trachea in front of the heart. Both this "tracheal lung" and the conventional lungs are simple vascular sacs with little internal division. Air moves in and out of the lungs by expansion and contraction of the rib cage; snakes do not have a diaphragm.

When a snake engulfs large prey, the mouth cavity is filled and the air pathway can be blocked. At such times a protrusible glottis (the opening and valve into the trachea) can extend outward to the edge of the mouth (usually to the side) beneath the food. So even with a full mouth, a snake can continue to breathe.

Breathing in snakes and other reptiles is a two-stage process consisting of a brief air-flow or ventilation cycle and a longer pause period. Constriction of the trunk forces the air out of the lungs (exhalation) and expansion of the trunk creates a semivacuum and draws air into the lungs (inhalation); then breathing stops. This breathing pause, or apnea, can last from a few seconds to several minutes in a resting and undisturbed snake.

In diving snakes, apnea is prolonged. Physiological tests on North American watersnakes (*Nerodia*) indicate that submergence lasts from 5 to 25 minutes while foraging and about 45 minutes while resting or sleeping. New data, more rigorously gathered, indicates that normal dive duration is actually shorter, averaging about 70 seconds and rarely exceeding 12 minutes (*Nerodia sipedon*). These observations also suggest that dive duration differs between species, by prey type, temperature, depth, and turbidity of the water, and the stimulus for diving.

The open-mouth threat display of the green parrotsnake (*Leptophis ahaetulla*) portrays this species as dangerous. It isn't. The protrusible glottis is visible on the floor of the mouth.

Seasnakes (Elapidae) submerge for 5 to 30 minutes, occasionally as long as an hour, while actively foraging. Most, and probably all, aquatic snakes take several breaths (ventilation cycles) during their brief moments at the surface. The duration of submergence probably depends on the depth of foraging. When foraging near the surface, snakes breathe frequently, once every 20 seconds to 2 minutes; the deeper they dive, the less frequently they surface to breathe. Similarly, a snake's level of activity dictates its need for oxygen. A snake that is actively searching or struggling with prey will need to surface more often.

Submergence times lengthen when an aquatic snake is sleeping or hibernating because metabolic activity slows. During such periods, cutaneous respiration (gas exchange through the skin) can provide sufficient oxygen and dispersion of carbon dioxide. Some individuals of the nonaquatic American ratsnakes (*Elaphe obsoleta*) were discovered hibernating underwater in a well, but it is not known whether they made periodic trips to the surface or actually remained underwater for the entire 4- to 5-month period of hibernation.

HOW DO SNAKES CRAWL AND SWIM?

Lack of limbs has not impeded snakes' mobility. Snakes move about in every imaginable medium except ice and snow. They crawl on and through soil and sand, over and among rocks, grass, and trees. They climb vertical rock faces and tree trunks, swim in streams, swamps, lakes, and oceans, and even glide through the air. (See Appendix 2 for crawling and swimming speeds.)

Undulatory Locomotion

Although snakes have evolved specialized locomotor behaviors for certain surfaces, most movement involves some type of serpentine or undulatory propulsion. The simplest mechanism is undulatory swimming. Muscular contractions, passing rearward (caudally), create a traveling body wave whose degree of curvature increases as the body becomes narrower toward the tail tip. Locomotor force arises from the outside rear edge of each curve pushing against the water. This lateral (sideward) and rearward push against the water propels the snake forward. The snake does not advance as fast as the body waves move backward because water offers little resistance to the snake's pushing and motion. To improve the effectiveness of locomotion in water, seakraits and seasnakes have evolved bodies that are compressed (flattened

The yellow-bellied seasnake (*Pelamis platurus*) is an excellent swimmer but nearly powerless when washed ashore. The absence of enlarged ventral scales, the compressed body, and the oar-like tail handicap seasnakes during terrestrial crawling.

Sidewinding, the North American sidewinder (*Crotalus cerastes*) leaves a discontinuous track in the sand because of its rolling-like locomotion.

side to side) and vertically compressed paddlelike tails. These flatter surfaces pushing against the water generate greater frictional resistance, a stronger propulsive force, and more forward movement with each undulation of body and tail.

A snake's locomotion on land is also undulatory but differs from its aquatic counterpart. Land surfaces are highly irregular and provide many frictional or resistant points. The curves of a snake's body moving on land are not regular. Instead, they match the irregularities of the surface, which the body pushes forward against. The forward thrust is localized on the outside rear edge of each body curve, and as the snake moves forward, the body-push arises continuously from more posterior edges of the body curves. Thus, on land a snake pushes simultaneously against the many surface irregularities along its body and the entire body moves forward at the same time. Each point from head to tail on one side of the body passes and usually pushes against the same resistant points. The tail follows the same pathway as the head and neck. Even when surfaces are discontinuous, such as the branches of a tree, a snake can use undulatory locomotion. Most snakes, except the highly specialized seasnakes, are capable of serpentine locomotion on land.

Sidewinding

A substrate of shifting sand is nearly always associated with sidewinding in snakes. Sidewinding combines aquatic serpentine locomotion with the terrestrial mode, but only in the limited sense that the body curves are uniform. Sidewinding depends on yielding resistant points, and the entire body follows the same pathway, on and off the sand. If we observe the movements of a sidewinding snake in slow motion, we see a series of uniform curves; one set of curve crests touches the sand and the other set is raised above the substrate. As the head and neck touch the sand, the area immediately below the neck also touches down. The remainder of the body then moves forward along this path. Meanwhile the head lifts off the substrate to form another raised curve, and the body subsequently follows along the same pathway. Obligate sidewinders are typically small snakes, and only one complete body curve is off the sand at any moment during sidewinding. Because of the raised body curves and yielding resistant surface, the track of a sidewinder appears as a discontinuous sequence of shapes at an angle to the direction of travel. Sidewinders reside in many deserts of the world and are not closely related to one another (for example, the sidewinder rattlesnake, *Crotalus cerastes*, of southwestern North America; the desert horned viper, *Cerastes cerastes*, of the Sahara and adjacent deserts; and Peringuey's desertadder, *Bitis peringueyi*, of the Namib Desert of southern Africa).

Slidepushing

Slidepushing was long misidentified as sidewinding in nonsidewinders. Like sidewinding, this behavior is characterized by uniform body curves and lateral undulation, but at a higher speed and with tighter curves than in sidewinding locomotion. Slidepushing allows for forward movement across near-frictionless surfaces such as slippery mudflats. It could be described as coordinated thrashing. The snake undulates rapidly, with its head held above the surface. The body curves pass backward, presumably twisting the sides of the body and engaging numerous scale edges against the slippery surface. The resulting friction, and the constant and widespread pushing, does not prevent the body from slipping but does generate enough frictional grab to move the snake forward—much more slowly than the backward-pushing curves of the snake's body. Indeed, without the constant high-speed push of rapid undulation, the snake would thrash without any forward movement.

Rectilinear (Caterpillar) Locomotion

The slowest and most methodical locomotor pattern is rectilinear, or caterpillar, locomotion. Using undulatory locomotion, a snake can crawl just as slowly but can readily shift to high speed. Rectilinear locomotion has two speeds, slow and very slow. This pattern is predominantly associated with slow stalking behavior in heavy-bodied snakes such as boas, pythons, vipers, and pitvipers. The movement largely involves shifting the body forward within the skin, using the muscle slips that link the vertebrae to the ribs and their respective ventral scales.

Each rib and associated scale are drawn forward by the contraction of muscle slips on the vertebrae immediately in front of the rib. When the rib and scale reach the limit of their forward movement, the muscle slips behind the rib contract. Because the scales have free rear edges, each rearward-contracted scale catches on surface irregularities, though body weight can provide enough frictional resistance. The vertebrae and associated body segments are thus pulled forward to the position of the anchored ventral scales. Waves of contraction pass rearward along the trunk, the ventral surface of the snake appearing to ripple forward and the body slowly but steadily shifting forward. This rippling effect, which mimics the movement of caterpillar legs, gives rectilinear locomotion its colloquial name. A snake that moves this way is also sometimes described as "walking on its ribs."

CAN SNAKES CRAWL BACKWARD?

Some legless animals like earthworms can crawl backward, but snakes cannot. A snake can turn the front half of its body rearward, change direction, and move headfirst in the opposite direction. A snake can also pull part of its body backward, but snakes cannot back up and move the entire body in reverse. The heavy-bodied rattlesnakes (*Crotalus*, *Sistrurus*) adopt a defensive posture in which the head always faces the enemy and the body slowly moves to one side or the other, but not backward.

HOW DO SNAKES CLIMB?

Snakes climb vertical surfaces using concertina locomotion. This locomotor behavior is a stop-and-go mechanism in which one portion of the body is stationary while another part moves. In concertina locomotion, the head and front part of the body extend forward until they reach an anchor point. Then the head and body form stationary curves while the rear part is drawn forward to the anchoring curves, where it forms a new set of stationary curves so that the fore part can be extended again. The process is repeated until the snake reaches its goal. This method of locomotion is also used when snakes traverse a smooth or low-friction surface.

Many snakes that live in trees move across branches using undulatory locomotion. To ensure adequate support, arboreal snakes must avoid wide gaps and thin branches. Where branches are numerous and reasonably horizontal, snakes can use undulatory locomotion as rapidly in trees as on the ground. Where branches are widely spaced or include many weak support points, however, stop-and-go concertina locomotion is preferable. First the snake anchors its hind end and slowly extends its head and upper body across the gap. When it establishes a new anchor

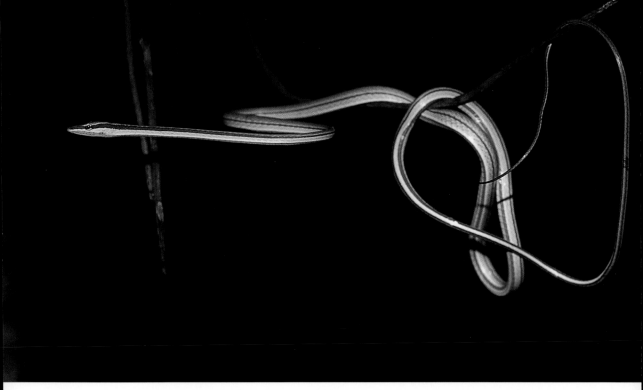

Living in trees requires the silver vinesnake (*Oxybelis argenteus*) to be capable of stretching across large gaps between branches and limbs.

point, it draws its rear part across the gap. Only rarely would a snake be able to extend half of its body length across such a gap. Specialists in this sort of arboreal locomotion include the chunkhead snakes (*Imantodes*), whose large heads are attached to long thin bodies (triangular in cross-section) that permit maximum body extension.

Climbing in a tubular animal such as a snake results potentially in major blood puddling or drainage as the snake changes its position from the horizontal. Circulatory adaptations have evolved that offset the lowering of arterial pressure to the brain (see *How Are Snakes Built?*). Additionally, a tight skin and relatively noncompliant tissue spaces oppose edema formation when a snake shifts its body from the horizontal.

CAN SNAKES FLY?

Only one genus, *Chrysopelea*, the flyingsnakes of Southeast Asia and the East Indies, is known to move through the air by gliding. It is not known whether all species of *Chrysopelea* glide.

Gliding is actually a controlled angular descent. By "sucking in its gut," the gliding snake increases the air resistance on its body, thus reducing the rate of fall. The sides of the ventral scales are rigid and the central part folds upward, producing a concave pocket along the full length of the trunk. This pocket acts somewhat like a parachute. In the best gliders, the resulting air resistance even produces lift. The snake hurls itself from its perch in the treetops, sucks in its ventral surface, and swims through the air with undulatory movements. Undulation aids balance, as does the tail, which is held rigidly upward and tilted back and forth like a tightrope walker using an umbrella for balance. The gliding snake has been observed to travel as much as 100 meters horizontally from its takeoff point. Although the snake's descent is gradual, it eventually lands with a distinct thud.

DO SNAKES DIG BURROWS?

Many species of snakes live below the surface, but few dig their own tunnels. Most depend on tunnels created by other burrowing creatures or cavities created by rotting roots. All subterranean snakes can push through soft soil for short distances to create an entrance to a natural tunnel or a temporary resting or egg-laying site. A few snakes, such as shieldtailed snakes (Uropeltidae), truly burrow, creating their own tunnels. Burrowing requires a sturdy skull and a spadelike or conical snout, because the head is a limbless animal's only possible digging tool. Digging is usually a modified form of concertina locomotion. The body forms a series of tight curves; the head pushes forward into the soil and lifts or moves from side to side to compact the soil to the roof or wall. This horizontal pile-driving behavior eventually creates a tunnel. Locomotion within tunnels is serpentine if the tunnel offers broad curves and surface irregularities; otherwise, concertina or rectilinear locomotion is used.

Some desert snakes habitually lie buried in sand but do not appear to crawl through the sand. Most appear to slowly wiggle beneath the sand by means of small side-to-side movements.

HOW DO SNAKES SEE?

Most, perhaps all, snakes have eyes, and their eyes are usually well developed. In predominantly subterranean snakes (Anomalepididae, Leptotyphlopidae, Typhlopidae, and Uropeltidae), the eyes are small and simplified; they often appear only as darkly pigmented spots beneath large head scales. Vision with such eyes is probably restricted to perception of light and dark. A few species, such as snouted blindsnakes (*Rhinotyphlops*), lack externally visible eyes, and these snakes may be truly blind. Snakes with well-developed eyes—small or large—register images, not just light and dark.

A life underground is often associated with the reduction or loss of eyes, such as these eye spots (left) in the western threadsnake (*Leptotyphlops humilis*). Above-ground life requires good vision and large eyes. Nocturnal species, like the horn-nosed sandviper (*Vipera ammodytes*) commonly have elliptical pupils (right).

The size and position of the eyes and the shape of the pupils are associated with habits and habitats. For instance, burrowing (fossorial) snakes and aquatic snakes have proportionately smaller eyes than terrestrial snakes and arboreal snakes. Diurnal snakes that use sight to hunt their prey often have pointed snouts and eyes positioned or aimed forward. This placement produces a wide overlapping field of vision directly in front of the head, and thus good depth perception. Ambush hunters and snakes that live in dense habitat-obscuring environments have more laterally placed eyes, which produce a broad visual field with less overlap. Some aquatic snakes, such as the wartsnakes (*Acrochordus*), have flattened heads and eyes positioned dorsally so they can see prey and predators above them.

Diurnal snakes (those active during the day) typically have round pupils and moderate-sized eyes. Nocturnal snakes typically have large eyes, and many also have vertical elliptical pupils; large eyes can gather more light in low-light conditions. A round pupil can close tighter to make a tiny opening permitting a minimum of light to enter the eye in very bright sunlight. In contrast, a vertical pupil can open wider to allow more light to enter the eye, an important feature for a snake that hunts at night in dim light. For such snakes, odor is likely to be as important as vision in locating prey, if not more important.

There is no evidence as yet that snakes have color vision. Vertebrate eyes have two basic types of light-receptor cells, rods and cones. Rods provide vision in low light, but do not yield sharp images; cones provide good visual acuity. Cones are also associated with color vision, which requires the presence of two or more types of cones with different visual pigments. Caenophidian snakes have three types of cones, but all three share the same visual pigment. Blindsnakes have only rods; boas and pythons have rods and only one type of cone cells.

WHAT ARE PIT ORGANS?

Eyes perceive light in the visible spectrum—that is, in the light frequencies visible to the human eye. Some snakes also "see" in the infrared spectrum using heat receptors called pit organs. Heat radiates or reflects from all objects as infrared radiation, and the frequency and intensity change with the temperature and mass of the object. Infrared vision detects a temperature difference between an object and its surroundings; basically, this difference produces a heat spot on a uniform-temperature background. Infrared vision is ideal for hunting warm-blooded (endothermic) prey, such as birds and mammals, at night. Indeed, many pitvipers, pythons, and boas feed predominantly on birds and mammals they hunt at night.

The infrared receptors usually lie in pockets or pits in the skin, which explains the name. Boas and pythons have many shallow pit organs along their upper lips and often on the lower lips as well. Several species of boas and pythons lack pits, but still have heat receptors beneath the scales of the snout. Pitvipers (Crotalinae), including rattlesnakes, have a deep pit on each side of the face between the nostril and eye. This crotaline pit contains a membrane, suspended above the bottom of the pit, that bears heat receptors, somewhat mimicking the function of the retina

Infrared reception evolved independently in several groups of snakes. The viperids (timber rattlesnake, *Crotalus horridus* [left]) have sensory cells in their loreal pits (between the nostril and the eye), whereas the boids (emerald treeboa, *Corallus caninus* [right]) have their infrared sensory cells in labial (lip) pits.

of the eye. Sensitivity to infrared radiation is 5 to 10 times higher in pitvipers than in boas and pythons. Some rattlesnakes (*Crotalus*) can detect a temperature change as slight as 0.003°C in 0.1 seconds.

Pit organs, whatever their number, are oriented forward. In pitvipers (and probably also in pythons and boas), the left and right fields of infrared vision overlap and provide precise directional information for aiming a strike.

Another light-wave receptor has recently been found on the tail of a species of tropical seasnake (*Aipysurus laevis*). The function of this receptor is unknown. It may serve to alert the snake that its tail is still exposed when the snake hides in crevices in coral. The discovery of this organ was entirely unexpected, and it is possible that other crevice-dwelling and burrowing snakes have them as well.

DO SNAKES HEAR?

As the old saying "as deaf as an adder" suggests, it is a common belief that snakes cannot hear. Snakes do not have external ears or eardrums, but they do hear airborne sounds. Sound reception differs from that of reptiles with eardrums: in snakes, sound is transmitted from skin to muscle to bone. Specifically, sound waves hitting the skin surface of the temporal area on the side of the skull are transferred through the jaw muscle to the quadrate bone. The ear bone (columella or stapes) abuts the quadrate, which is loosely attached to the underside of the cranium, and in turn transmits minute vibrations from it to the inner ear and its sound-sensitive cells, exactly as in other reptiles. This skin–muscle–bone transmission route is most effective for low-frequency sounds such as seismic vibrations, which are typically less than 200 hertz, and most snakes hear best in the 150- to 600-hertz range with a peak response at 300 hertz. Although both the muscles and quadrate are part of the jaw mechanism, there appears to be no degradation of sound when a snake is eating.

Early theories about snake hearing hypothesized poor reception of low-frequency sounds only through the lower jaw when it was in contact with a surface. Snakes can sense vibrations in this way, but the reception is more tactile than auditory. Furthermore, snakes hear via the columella whether or not the head touches a surface. A recent experiment on a single individual of one species suggests that the lungs also act as receptors when sound strikes the body wall. It is not known how sound vibrations received by the lungs are transmitted to the inner ear.

CAN SNAKES SMELL?

Odor detection is a critical sense for all species of snakes. It is used to find food and mates, to detect predators, to stimulate courtship, and for other activities.

A Javan red-tailed ratsnake (*Gonyosoma oxycephalum*) is tongue flicking to find odors.

Snakes' tongues, which are seldom still, play an important role in this sense. The tongue flicks in and out through a notch in the front of the upper lip, and the flicking becomes more frequent when a snake is exploring. When extended, the tongue tip waves up and down and odor particles adhere to it. When the tongue is withdrawn into the mouth, the odor particles are transferred to the roof of the mouth near a pair of duct openings leading into a special olfactory chamber, the Jacobson's or vomeronasal organ, which lies below a larger olfactory chamber.

We are still uncertain how the particles move from the tongue into the fluid-filled Jacobson's organ. Experiments using carbon dust and radioactive particles have demonstrated that particle transfer from the tongue to Jacobson's organ does occur. We also know that the tongue tip is not inserted into the vomeronasal organs, as was long supposed. It appears that particles are transferred from the tongue to a tissue pad on the floor of the mouth, and that this pad is then pressed against the roof of the mouth, transferring odor particles to the Jacobson's organs. Thus, the tongue does not smell—that is, it does not actually sense the presence of odors—but it is necessary for carrying odor particles.

Snakes also smell through their noses. Airborne odors are sniffed or breathed in through the nasal passages (also paired, as are the vomeronasal organs) into the olfactory chambers. Both the olfactory chambers and the Jacobson's organs have patches of sensory cells, and the chemical reaction of odor particles with the cell surfaces is registered as nerve impulses, which are sent to the brain for interpretation.

The sense of smell is of particular importance to snakes in tracking the paths of other snakes and of prey—for instance, males need to track females, and vipers must first envenomate and then follow their dying prey. A snake's locomotion involves pushing against objects, which then bear odor signatures on only one side, the side pushed against. When a second snake passes the same way, one of its tongue tips picks up more odor particles than the other tip, indicating whether the left or right surface of a series of objects was touched by the earlier snake. Thus the tracking snake can identify the first snake's direction.

DO SNAKES TASTE THEIR FOOD?

Turtles and most lizards have many taste receptors on their tongues and other oral surfaces. Snakes do not. Modification of snake tongues for transport of odor particles apparently has eliminated taste receptors on the free part of the tongue, and probably from the entire tongue. Reports of taste buds or similar structures in thread wormsnakes (Leptotyphlopidae) and seasnakes (Elapidae) appears to result from researchers' incorrect interpretation of some microscopic structures on the epithelium of the mouth. A broader survey of the oral epithelium in snakes did not reveal taste buds.

HOW DO SNAKES FIND AND SUBDUE THEIR PREY?

Some snakes search for their prey. Others sit and wait for prey to come to them. Both hunters and ambushers use sight, smell, hearing, and touch to locate, recognize, and capture their prey. One sense may predominate, but it is unlikely that any species of snake depends exclusively on a single sense. A good example of the use of multiple senses is the ambushing technique of the timber rattlesnake (*Crotalus horridus*) of eastern North America. The rattlesnake forages across the forest floor until it locates a rodent pathway by smell. The snake then follows the trail until it sees a suitable ambush site where the path runs along the top of a fallen log. Positioning itself beside the log, the rattlesnake rests its chin on the log, perpendicular to the path, and waits. When a mouse scurries onto the log, its arrival is detected by vibrations (hearing and touch). If there is sufficient light, the snake also sees the mouse with its eyes. As the mouse passes in front of the snake, the pit organs and eyes aim the strike. Contact with the mouse's body triggers the injection of venom and brief closure of the mouth around the mouse before the snake recoils. The mouse continues to run but soon collapses from the shock of envenomation. The rattlesnake uses smell, mainly from tongue flicks, to track and find the mouse.

Waiting to ambush its prey, a timber rattlesnake (*Crotalus horridus*) rests its head adjacent to a regularly used rodent trail along the log. (T. Borg, courtesy of Savannah River Ecology Laboratory)

This interaction of multiple senses characterizes all snakes. In one instance, an individual snake might depend mainly on sight to locate and capture its prey; at another time it might rely on smell. A snake is not a mere stimulus-response mechanism. As circumstances, experience, and prey opportunities change, a snake will adjust its hunting behavior to improve its success. Like all predatory animals, however, each species of snake hunts a specific set of prey and uses specific hunting behaviors.

Finding prey involves chance or luck—being in the same place at the same time as the prey—but hunting is not random. When western rattlesnakes (*Crotalus viridis*) leave their wintering dens and disperse over the adjacent countryside, each follows a fairly straight route and stops when it finds an area with many rodent-scent trails; many trails suggest many rodents. The rattlesnake selects a good ambush site and begins its vigil. After several weeks of successful hunting, prey encounters fall to an unprofitable level, and the snake moves on to a new hunting ground. Watersnakes (*Nerodia sipedon*) patrol the water's edge on warm, rainy summer evenings, where and when they are most likely to find frogs. Such productive hunting routines are used by all snakes.

Learning fine-tunes feeding instincts; snakes learn where, how, and for what to search. Hatchling snakes of the same clutch will fix on different prey if their first meals are different. Experiments show that newborn gartersnakes (*Thamnophis*) that taste earthworms first will subsequently prefer earthworms, and those that first taste frogs will prefer frogs. These preferences then give rise to different hunting behaviors.

How Is Venom Used in Prey Capture?

A venomous bite slows or stops a prey's escape by disrupting normal body functions. This is how snake venoms work now, but we do not know whether venom evolved as an adaptation to capture prey or to speed digestion of the prey once it was swallowed (see *What Is Venom?*).

No matter which hypothesis is correct—or both may be—venom is effective in prey capture and many venoms begin digestion (particularly in Viperidae). Fixed-fang snakes (rear-fanged colubrids and front-fanged elapids) usually hold their bite, sometimes throwing body loops over or around the prey, but not constricting. Chewing action then injects the venom into the muscles or the body cavity. In contrast, the hinged-fang snakes, such as viperids and molevipers (*Atractaspis*) and deathadders (*Acanthophis*), usually strike-bite, simultaneously injecting venom into the prey, and then allow the prey to stagger away—later to be tracked and eaten (see *How Do Fangs Work?*). Both methods minimize struggle between the snake and its prey and thus apparently reduce the possibility of injury to the snake. Some rattlesnakes bear facial scars, probably inflicted at the moment of the bite by larger prey such as chipmunks, squirrels, and woodrats.

How Does Constriction Kill?

Constricting snakes do not simply throw body coils around their prey. They also tighten—constrict—these coils with each of the prey's exhalations and struggle movements. The association of constriction with exhalation suggests suffocation—that is, preventing the prey from breathing, causing its asphyxiation and death. Careful observations often indicate, however, that the prey has not been constricted long enough to cause its death by suffocation. Instead, the coils around the prey's chest collapse the chest cavity, deflating the lungs and compressing the heart. Compression of the heart prevents normal pumping and leads to immediate circulatory arrest and insufficient blood flow to the heart muscles, brain stem, and other major organs. Cell damage in these critical organs is immediate, and death follows soon thereafter. Constriction can crack a rib or other small bones, but such breaks are rare and the prey's skeleton is not broken, crushed, or disarticulated. Constricting snakes often specialize in bird and mammal prey.

A red milksnake (*Lampropeltis triangulum syspila*) constricts its prey (skink, *Eumeces*) by completely encasing it in several body loops. (R. W. Van Devender)

Some constrictors hunt by searching and others hunt by ambush, but in all cases the first capture movement is a strike-bite. The bite locks the snake to the prey. If the prey is small, it is pulled back toward the snake and quickly encircled in body coils. Large prey is anchored by its own weight while the snake pulls its own body toward the prey and then coils around it. In both situations, the bite and encircling occur in a blur of movement. Once it has encircled its prey, the snake might adjust its bite or retain the original bite-hold until the prey dies.

Do All Snakes Kill Their Prey before Eating It?
Venom injection and constriction both kill the prey. These methods of capture are commonly associated in nature with large prey that cannot be swallowed unless it is immobilized. After the prey's death, the snake swallows the prey headfirst, compressing the body for efficient swallowing. Otherwise, one of the prey's legs might block the snake's throat, preventing either regurgitation or swallowing.

However, many snakes do not feed on large prey. For these snakes, capture, eating, and swallowing are a continuous process. The bite secures the prey and brings it partly or totally into the mouth. Once the prey is in the mouth, the jaws "walk over" it, and it is swallowed alive.

WHAT AND HOW DO SNAKES EAT?

Snakes eat only animals. Many species of snakes (generalists) eat multiple species, although the prey may be limited to a particular group of animals such as arthropods. Other species (specialists) eat only one or a few types of prey. In some cases, prey availability determines whether a snake species in a particular locale is a specialist or generalist.

Because snakes have no limbs with which to hold prey or tear it into fragments, the mouth and teeth must hold the prey and then move it into the digestive tract. Because the tongue is slender and used for odor detection, it cannot assist much, if at all, in food movement.

The mobility of the jaw and mouth skeleton enables snakes to swallow large food items. The left and right sides of the jaws move independently and the mouth can open wide, both crucial for moving the food item from the mouth cavity into the esophagus.

To understand how a snake eats, it is necessary to understand the skeleton of the mouth and jaws. The lateral view in Figure 1.1 shows the right side of a snake's head. The bones it shows are matched on the left side. Four bones form a free-floating upper jaw, attached to the braincase and quadrate bone by tendons, ligaments, and muscles. Two of the bones, the palatine and the pterygoid, lie tip-to-tip on the roof of the mouth beneath the braincase (see the ventral view in Figure 1.1; together they extend the full length of the mouth from the snout to the esophageal opening. A third bone, the maxillary, forms the outer edge of the mouth and extends from the snout to behind the eye. These three bones all bear teeth. The fourth bone, the toothless ectopterygoid, connects the internal palatine-pterygoid arch to the maxillary bone in the rear of the mouth.

The lower jaw is a linear series of bones; only the frontmost one (the dentary) bears teeth. At its back end, the lower jaw joins the skull via a hinge joint with the quadrate bone in the upper jaw. The quadrate in turn joins the braincase via a moveable attachment, which permits the jaws to move in or out and forward or backward. This motion allows the mouth cavity to widen enough to engulf prey larger than the snake's head.

As the jaws on one side of the head begin to open, those on the other side bite down. The opened jaws then shift forward, disengaging their teeth from the prey and biting it anew. As one side bites, the other side shifts forward. Thus the left and right sides ratchet over the prey, forcing it rearward into the esophagus.

Although the shape and size of snakes' teeth differ among species, they are typically elongate rearward-curving cones with sharp-pointed tips. The size and shape of teeth also differ among the different jaw bones and by location on a particular bone. Snakes' teeth are grasping tools, and their rearward curve ensures that they easily disengage only as the food moves toward the rear of the mouth.

All snakes are predators. Most small snakes, like this Yaquia black-headed snake (*Tantilla yaquia*), eat invertebrates, from snails and slugs to a full variety of arthropods, even venomous ones such as centipedes.

When a snake strikes, the mouth is often closed or barely open at the beginning of the strike, but fully open (in some cases, the upper and lower jaws nearly form a 180-degree angle) as the mouth reaches the prey. The curved teeth enter nearly straight into the prey, and the rapid shutting of the snake's mouth brings the prey's body, or part of it, deep into the mouth cavity. The prey's struggle only serves to deepen the penetration of the teeth and to tighten the snake's oral grasp. Swallowing can begin immediately with small, struggling prey and after larger prey is subdued by constriction or envenomation. When eating subdued prey, the snake usually releases its bite and moves its own head to the head of the prey. The independent ratchetlike movements of the left and right sides then move the jaws over the prey. The prey slowly but persistently shifts rearward into the esophagus, receiving a copious coating of saliva. Once the prey passes beyond the teeth, the neck and trunk muscles twist and flex the snake's body; body loops shove the prey rearward toward the stomach. Once the prey is safely in the esophagus, the snake performs a series of yawnlike manipulations of the mouth. These yawns realign and reset the mouth and jaw bones in preparation for the next feeding strike.

This eating-swallowing mechanism characterizes most snakes but is modified in various ways for feeding on special prey. The South American snail-eating snakes (*Dipsas*) can extract a snail's body from its shell. *Dipsas* first bites the exposed part of the snail. This bite prevents the snail from withdrawing into its shell and allows the snake to orient its head directly in front of the shell opening. The lower jaw is thrust between the shell and the snail's body. Retraction of the lower jaw and ratchetlike movements of the upper jaw then remove the snail from its shell, usually within 2 or 3 minutes. The North American brownsnake (*Storeria dekayi*) also eats

A Texas long-nosed snake (*Rhinocheilus lecontei*) has captured an ornate treelizard.

snails, but twists them out of the shell instead of "sucking." The brownsnake bites the body of the snail and pushes the snail forward until its shell becomes wedged against some object. Then the snake rolls, usually clockwise. After 10 to 15 minutes of such twisting pressure, the snail's muscles fatigue and relax, and the snake tears the body from the shell.

Blindsnakes are ant and termite specialists and their jaw mechanisms permit them the rapid intake of eggs and larvae, thus reducing their exposure to the attacks of adult termites and ants. Blindsnakes (Typhlopidae) use an alternating left and right extension of their upper jaws to drag prey into their mouths. In contrast, threadsnakes (Leptotyphlopidae) use a bilateral mandibular raking mechanism that involves the synchronous outward and inward movement of the lower jaw.

The African egg-eating snake (*Dasypeltis scabra*) feeds exclusively on bird eggs. Such a diet requires (1) an eating-swallowing mechanism that can engulf smooth-surfaced objects two to three times larger than the snake's head, and (2) a mechanism that rapidly disposes of the shell and digests its contents. During the birds' brief nesting season, the snake must obtain sufficient energy to last throughout the 8- to 9-month interval when no eggs are available. When it finds an egg, the egg-eating snake crawls around it, presses its open mouth against the end of the egg, and pushes. The egg slides against the snake's body and stops; the snake's mouth begins to stretch over the egg, and the pushing forces the egg totally into the mouth cavity. Then peristalsis-like contractions of the chin and throat muscles move the egg into the esophagus, where it stops. Muscle contractions of the neck and tilting head drive large, sharp vertebral spines into the egg and crack it. Continuing compression of the egg empties its contents into the snake's esophagus and stomach. When the egg is empty, the snake regurgitates the flattened eggshell and is immediately ready to eat another egg. Many other species of snakes eat bird eggs, but only a few have long and sharp vertebral spines to break the egg, and none of the others regurgitates the shell. Egg-eating snakes that cannot break the shell must carry the egg around as a body lump until digestion dissolves it.

The North American hognose snakes (*Heterodon*) are toad-eating specialists. Physiologically resistant to toad-skin toxins, they also have enlarged swordlike teeth on the rear of the maxillaries. The swallowing mechanism is identical to that of other snakes, but toads puff up their bodies in defense. The snake's enlarged rear teeth were once mistakenly assumed to penetrate the toad's body wall and puncture its lungs, deflating the toad like a punctured tire. Experiments and behavioral observations indicate, however, that the teeth are too short to deflate the toad; instead they channel venom into the bite, and envenomation relaxes the toad for ease of swallowing.

WHY AND HOW DO SNAKES SHED THEIR SKIN?

Like that of other reptiles, a snake's skin has two layers, an inner dermis and a thinner outer epidermis. The dermis contains the blood vessels, nerve endings, glands, and connective tissue. The epidermis is a protective sheath that shields the dermis and all other organs from abrasion and puncture through its rows of overlapping scales. The scales are produced by the epidermis, unlike those of fishes that are produced by the dermis. The epidermis and scales form a physical and physiological barrier against bacteria and external parasites, regulate body temperatures, water balance, and gas exchange, and produce sex-related pheromones.

As the major line of defense against environmental hazards, the skin is exposed to many traumas, such as scrapes, cuts, blows, and bites. As a result, it wears out and its outer layers must be replaced periodically. Shedding and replacement involve only the epidermis. The base of the epidermis (stratum germinativum) periodically produces new cells through mitosis (Figure 1.2). This process creates new layers of cells between the old outer skin, or sheath, and the newly produced inner sheath above the stratum germinativum. As the surface cells of the new sheath age, their

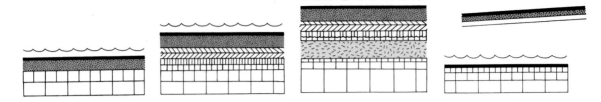

Figure 1.2. Major stages in the sequence of new skin growth and shedding. (A) During the resting phase, cell division and multiplication is extremely slow. (B) Then an active growth phase begins, during which cells multiply. (C) When the active phase ends, the cells between the old outer skin and the new skin break down and liquify. (D) Eventually the old skin separates from the new skin and is shed. (Adapted from Landmann 1986)

walls become increasingly thick and keratinous, and they eventually die. The resulting layer of dead cells forms a dense keratinous cover, capped by the Oberhautchen, the outermost portion of the epidermal sheath.

The technical term for the process of skin-shedding or molting is *ecdysis*. How often ecdysis occurs depends on many factors, but particularly on how well the snake has fed and therefore how much it has grown and on how much damage the skin has experienced. Ecdysis thus does not occur on an invariant schedule. The cycle varies between individual snakes and even within the life of an individual, occurring more frequently in faster-growing juveniles than in adults. Humans, by contrast, continually shed individual skin cells.

Shedding can occur in any season when a snake is active; among tropical snakes that are active year-round, shedding occurs throughout the year. It does not appear to occur during winter hibernation or in summer among snakes that become torpid in hot weather. Most newborn snakes shed their skin within 24 to 36 hours of hatching or birth, although in some individuals, ecdysis does not begin for about a month. In snakes with skin injuries, the ecdysis cycle speeds up and the number of shedding events increases to heal wounds.

Ecdysis begins with the death and liquidization of the bottom cells of the old outer sheath, above the new sheath's Oberhautchen (Figure 1.2). These dissolving cells cause the outer skin to soften and become dull. The scale (spectacle) covering each eye becomes milky and opaque, temporarily clouding the snake's vision. In such a state, snakes stop feeding and remain hidden. As a defensive measure, they are often ill-tempered, and even mild-mannered snakes frequently strike at such times. In a few days, however, the spectacles clear, and within 4 or 5 days, the snake sheds the old skin sheath in a single piece. Shedding usually begins at the lip (labial) scales, as the snake rubs its face against rough objects to fold the labial scales back on themselves. Once the layer of scales over the entire head is free, the

A Caucasian viper (*Vipera kaznakovi*) has just begun to shed.

snake repeats the rubbing process; catching the loose skin on a rock or branch, the snake crawls forward out of it. Crawling out of the old skin turns it inside out, much like peeling off a sweater. Eventually the old skin pulls free from the tail. The castoff skin, thin and transparent, is abandoned. Unlike many lizards, snakes do not eat their newly shed skin.

The shedding cycle typically takes about 14 days. The new skin is more boldly colored and vivid in pattern. After a resting phase that can last from a few days to several months, the next shedding cycle begins.

WHY ARE SNAKES STRIPED, BANDED, OR BLOTCHED?

Snakes rely on their coloration and patterns to hide from both predators and prey. Avoidance is the safest and least effortful defense against predators, and visual mimicry of a nonprey item enables the snake to disguise its identity and dupe potential predators. When it is not recognized as food, the snake avoids attack and the need to defend itself. A snake's appearance can mimic a particular nonprey item or the general background, or its pattern can break up the snake's outline on a varied background.

Irregular blotches, regular spots, transverse bands of alternating colors, and other contrasting patterns obscure the outline of a snake, allowing it to blend in with an

A boldly patterned North American copperhead (*Agkistrodon contortrix*) is well camouflaged in leaves. The graded colors between the snake's blotches match the surface of fallen leaves and hide the snake.

irregularly colored background. Visual disruption of the body's outline succeeds whether the snake is stationary or moving. Boa constrictors (*Boa constrictor*), Gaboon vipers (*Bitis gabonica*), and American copperheads (*Agkistrodon contortrix*) are boldly patterned and live in forest areas with heavy leaf litter. The colors and irregularities of these snakes' patterns make them almost invisible while lying still among leaves. Their camouflage not only protects them but also hides them from prey animals, and all three snakes routinely use ambush to capture prey.

Stripes appear to be an effective camouflage when a snake is moving through thick vegetation. Although visible, the snake's continuous line disguises its movement—and then suddenly the snake is gone. Most striped species are slender and fast-moving.

Bands of alternating color create an impression of flickering light and dark when a snake is moving. This flickering effect makes the direction of movement confusing to the viewer. When the snake is moving fast, it can also create the impression of a unicolored snake that seems to disappear when it stops moving. Banded snakes live in a variety of habitats from forest to desert. Many either burrow or spend much of their time beneath surface litter, and their alternating bands disrupt the body outline when part of the snake is exposed.

More uniformly colored snakes are visually hidden by the close match between their green, brown, and gray skins and the colors of their habitats. The body colors of many desert species match the particular sands, soils, or gravel of their environments. Sidewinder rattlesnakes (*Crotalus cerastes*) and other desert snakes often differ in color by locale, ranging from nearly white on white sands to beige or reddish on sands of those colors. Many arboreal snakes, such as green mambas (*Dendroaspis*) and American greensnakes (*Opheodrys*), match the foliage of the bushes and trees through which they forage.

The bright colors and bold patterns of other snakes advertise their presence. The message of their appearance is a warning: "I am dangerous" or "I taste awful." Such warning advertisement, called *aposematism*, is widespread throughout the animal world. Aposematism has led to the independent evolution of mimicry complexes whereby a harmless or tasty species (the mimic) matches a dangerous or distasteful species (the model) in color and pattern.

Coralsnake mimicry is the best-known mimicry complex in snakes. Coralsnakes have alternating bands of two or three colors. The bands are either sharply contrasting dark and light or very bright blacks, reds, and yellows. Within the Americas, there are more than 50 species of venomous coralsnakes (Elapidae: *Micruroides*, *Micrurus*). Their color patterns range from black-and-red-banded species to the more typical black-yellow-red-banded coralsnakes. (Remember, the old adage "Red and black, venom lack / Red and yellow kill a fellow" applies only in the United States!) In most areas with venomous coralsnakes, there are also one or more non-venomous or mildly venomous banded snakes (for instance, Colubridae: *Atractus*, *Erythrolamprus*, *Lampropeltis*, *Urotheca*) with nearly identical color and banding patterns. The close resemblance of venomous and nonvenomous species in the same area supports the hypothesis that the nonvenomous are truly mimicking the venomous species. The behavior of some snake-eating birds attests to the effectiveness of coralsnake mimicry. The young of some tropical snake-eating birds instinctively avoid any long thin object with a red-yellow-black-banded pattern, and some scavengers (mostly birds) that consume road-killed snakes typically leave dead coralsnakes and their mimics alone.

Mimicry of venomous snakes occurs in other forms as well. The triangular head, flattened body, hissing, and C-shaped striking coil of vipers are mimicked by a variety of nonvenomous colubrid snakes the world over. For instance, the toothless egg-eating snake *Dasypeltis scabra* of Africa not only matches the saw-scaled viper (*Echis carinatus*) in color, pattern, and defensive behavior, but also creates the same rasping sound by rubbing together the scales on adjacent body loops.

Some snakes, such as the American copperhead and Australian deathadders, use another form of mimicry to attract and capture prey. In juveniles of these species,

Mimicry requires a model, the venomous harlequin coralsnake (*Micrurus fulvius* [left]), that is distasteful or venomous so that the mimic, the scarlet kingsnake (*Lampropeltis triangulum elapsoides* [right]), is less likely to be attacked by a predator that has learned to avoid the model.

the tip of the tail is uniformly yellow and highly visible, unlike the cryptically colored body. The tail is held upright and waved, resembling a wiggling worm. When a curious lizard or other prey approaches to grab the "worm," it is seized by the luring snake.

How Do Snakes Get Their Colors?

Snakes display the full spectrum of colors, from white to black, from red to blue. Some species are uniform in color; others bear multicolored bands, stripes, spots, speckles, or mottling. Smooth scales often appear shiny or glossy; keeled scales (those with a raised ridge down the center) usually appear dull.

All colors except blue arise from color cells (chromatophores) embedded in the innermost layer (dermis) of the skin (Figure 1.3). There are four main types of chromatophores. Melanophores contain melanin and can produce blacks, browns, yellows, or reds, depending on the density of melanin and the abundance and position of melanophores in the dermis. Melanin, present in the skin of humans and most other vertebrates, is the dominant pigment in snakes. Xanthophores contain carotenoids and pteridines, producing shades of yellow or orange. Erythrophores also contain carotenoids and pteridines, yielding shades of red and orange. Iridophores contain purines that are colorless but reflective and often iridescent. Some shiny-scaled snakes appear iridescent because of the play of light on the scales' outer surfaces or on the buried iridophores. Blue results from a differential scattering of light from scale surfaces and subsurfaces. The various chromatophores are stacked, and different arrangements produce the range of colors seen in snakes. For example, green treesnakes appear green because the blue of scattered light com-

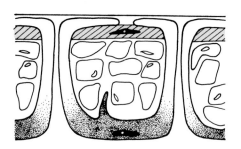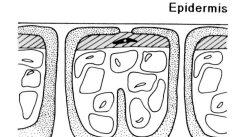

Figure 1.3. Chromatophore stacks in reptilian skin (schematic cross-section). Each chromatophore unit consists of a melanophore (stippled cell) and several layers of iridophores (clear cells) lying beneath a flat, lenslike xanthophore (diagonal-lined cell). These units are located in the dermis immediately below the epidermis. (Left) When pigment is concentrated in the bottom of the melanophore, the skin is light. (Right) When pigment spreads throughout the melanophore, the skin is dark. (Adapted from Taylor and Hadley 1970)

bines visually with the yellow of xanthophores to produce green. The yellow pigment, which is a fat-based compound, breaks down when the snake dies, and the skin slowly changes from green to blue or bluish-grey.

When Is an Albino Not White?

White represents the total reflection of light, black the total absorption of light. The colors of vertebrates, including snakes, depend on the presence, distribution, and abundance of melanin (see Figure 1.3). When melanin is spread throughout the melanophores, the animal appears dark. If the melanin is concentrated in a single location within the cells, the animal appears light. White bands or spots on snakes result from the absence of melanophores, erythrophores, and xanthophores, but the presence of reflective iridophores.

Albinos lack melanin. Absence of this dominant pigment makes the skin translucent, so that the red respiratory pigments (hemoglobin) of the blood show through, particularly in the eyes, tongue, and other thin-skinned surfaces. Absence of melanin is a genetic mutation.

Albino snakes are predominantly white, but the spots, stripes, and other markings formed by xanthophores and erythrophores can persist in shades of orange to red, usually pale due to the absence of melanin. "Partial" albinos are called xanthic albinos when the yellow-orange markings persist and erythristic albinos when red-orange markings persist. In other genetic mutations, melanin is present but yellow-orange or red-orange pigments are absent. For example, anerythristic cornsnakes (*Elaphe guttata*) have dark-brown spots on a grayish-brown background instead of a bright reddish-orange pattern.

"Total" albinism—absence of all pigments—is rare. Such a snake would be patternless and white, except for pink eyes and other thin-skinned surfaces. (Survival with no pigments in the blood is impossible because these pigments carry oxygen to cells and tissue.) Equally rare are leucistic snakes, which are white-skinned and patternless but have normally pigmented eyes. Their skin lacks all pigments, but melanin is present in the retinas.

Can Snakes Change Colors?

Few snakes are capable of quickly adjusting their colors to their backgrounds. Only the western rattlesnake (*Crotalus viridis*) has been reliably reported to change quickly (1 to 2 minutes) from light to dark or the reverse. Other species that may also be capable of rapid darkening and lightening have not been studied. Slow color change has also been inadequately studied. A few species, such as dwarfboas (*Tropidophis*) and Madagascan boas (*Sanzinia*), darken and lighten in response to activity or the daily cycle of dark and light. Their color changes take 60 to 90 minutes.

Some species change color and/or pattern with age or season. The eastern black ratsnake (*Elaphe obsoleta obsoleta*) and the black racer (*Coluber constrictor constrictor*) have distinct juvenile patterns of dorsal blotches and small dark spots on a lighter background; as they mature, the pattern disappears due to progressive blackening (melanization) of the skin, finally producing black adult snakes. Similar juvenile-to-adult color changes occur in other species, although such marked changes are not common. Even fewer species display seasonal color changes, although some males become brighter during the breeding season (for example, the European common viper, *Vipera berus*, brightens after spring shedding and before beginning courtship).

HOW DO SNAKES GROW?

Like all vertebrates, snakes grow by expansion—that is, by enlarging each muscle, bone, or scale through the development of new tissue. Snakes do not grow longer and larger by adding new scale rows or new vertebrae. Snakes grow between shedding events; they do not wait for the shedding cycle to expand in size. Shedding is a natural and necessary process, but—unlike the rigid exoskeletons of insects and crustaceans—snake skin is elastic, growing and stretching naturally as the underlying bones and muscles grow larger.

Do Snakes Grow throughout Their Lives?

Some species of snakes never stop growing, although growth in length becomes nearly imperceptible in older snakes. The general pattern for reptiles, including snakes, is one of rapid growth until sexual maturity; growth then slows significantly, particularly in females, which must use most of their energy reserves for egg production. Nevertheless, observable growth continues. How long this slow growth continues depends on the particular species.

Growth throughout an animal's entire life is called *indeterminate growth*. The pattern of juvenile and young-adult growth followed by no further adult growth is known as *determinate growth*. Until recently, indeterminate growth was attributed to all species of snakes, but biologists are no longer certain. The narrow range of body lengths in many smaller-bodied species suggests that growth might stop completely after several years of slow adult growth. Confirmation would require following the same individuals in the wild and periodically recording their body lengths. Aside from the difficulty of accurately measuring a live snake, very few small snake species have been tracked for the decade or more that would be required to confirm determinate growth.

How Fast Do Snakes Grow?

Speed of growth can be measured in two ways: by how long it takes for a snake to reach sexual maturity (adulthood), and by how much its length increases in a specific interval of time (such as centimeters per month).

Some snakes mature in a year or less; others require 2, 3, or even 9 years (female timber rattlesnakes, *Crotalus horridus*). Age at maturity is broadly associated with adult body size and climate: by and large, small to medium-sized species mature faster (1 to 2 years) than larger species (2 to 4 years). Age at maturity also often differs between males and females; the usual pattern is for males to mature earlier and at a smaller size.

It is more common to measure growth by examining change in body size per unit of time. Relative growth rates can then be compared at different periods in an animal's life, in different seasons, between the two sexes, or among different populations or species. The first year of life is the period of greatest growth. Many species double or nearly triple their length then. If several more years pass before maturity, growth will subsequently slow somewhat, but not greatly, after the first year. If maturity occurs during the first year, or shortly thereafter, a snake's growth rate will begin to slow gradually during the first year.

Many other factors besides age affect a snake's rate of growth. Availability of prey and water, and quality of food, are major factors. If prey and water are abundant, growth is rapid. Similarly, when the quality of food is high (high in proteins and nutrients and easily digestible), growth is fast. Conversely, poor-quality or insufficient food slows or halts growth.

The ability to stop growing temporarily during food shortages is one way in which snakes and other reptiles are superior to mammals. The low metabolism of snakes allows them to survive without food for extended periods. A mammal might also stop growing, but because of its higher metabolism it will starve if the food shortage lasts too long. Once a mammal stops growing, furthermore, it cannot regain the vigorous rate of growth it enjoyed before the bout of starvation and will ultimately be smaller than it would otherwise have been. A snake, by contrast, resumes growing at approximately the same pace as before.

Temperature also affects the growth of snakes and other reptiles. Snakes are ectotherms (see *What Does It Mean to Be Cold-Blooded?*), which means that they depend on external heat to raise and maintain an elevated body temperature. If body temperature is too low—less than approximately 16°C—growth will stop, no matter how abundant the prey, because the snake cannot effectively digest its food. As body temperature rises, digestive efficiency and other body functions improve and growth rates increase. For this reason, all temperate-zone and many subtropical snakes grow cyclically: they grow during spring and summer and stop growing in fall and winter.

Even though large-bodied species are genetically programmed to grow large and small species to stay small, different individuals have different growth rates and reach different adult sizes. These differences would persist even if two individuals ate identical foods and experienced the same environment. Metabolism and digestive efficiency differ genetically, as do other physiological processes.

Finally, a snake's health affects its growth rate and adult size. An injury or illness during juvenile growth can permanently modify an individual's growth potential.

HOW LONG DO SNAKES LIVE?

The oldest known snake was a ball python (*Python regius*) that lived 47 years at the Philadelphia Zoo after arriving there as a young adult. Life spans of 15 to 30 years are not unusual for large snakes in captivity. In the wild, the hazards of life usually result in much shorter life spans. How much shorter is difficult to predict because few snake populations have been monitored for more than a few years.

Snakes lack external traits that reveal their ages. Growth in snake skeletons does not match the annual growth cycle, although the number of growth layers in bones might reflect the number of times a snake has shed plus the number of annual growth cycles. As well, within a species or population of snakes, the largest individual is probably not the oldest but merely the fastest-growing. The only foolproof method of determining an individual snake's age is to mark it at birth or hatching and periodically recapture it until it dies. Without long-term ecological studies, tracking many individuals in this way is impossible.

TABLE 1.1. NATURAL LONGEVITY: LIFE SPANS OF FREE-LIVING (WILD) SNAKES

Species	Maximum age (years)	Adult size[a] (centimeters)	Age at first reproduction (years)	
			Female	Male
Timber rattlesnake				
(*Crotalus horridus*)	≥25	(90–100)	7–8	—
European asp				
(*Vipera aspis*)	18	45–50	5–6	4–5
American copperhead				
(*Agkistrodon contortrix*)	18	51–68	3–4	3
Ringneck snake				
(*Diadophis punctatus*)	≥17	24–26	2.8	2.8
Olive seasnake				
(*Aipysurus laevis*)	15	(100–190)	4–5	3
Eastern racer				
(*Coluber constrictor*)	9.1	70–100	2	2
Common gartersnake				
(*Thamnophis sirtalis*)	9	60–80	2	2
Rough greensnake				
(*Opheodrys aestivus*)	8	40–45	1.8–2.8	1.8
Japanese four-lined snake				
(*Elaphe quadrivirigata*)	7.8	(100–110)	3	—
Oriental tigersnake				
(*Rhabdophis tigrina*)	6.1	(90–100)	1.3	1.3

Notes: Various ecological techniques were used to obtain age data. In some cases, age at first capture was estimated when determining maximum life span. Long dashes signify lack of scientifically valid data.

[a]Average snout–vent length of adult females. Sizes in parentheses represent total length.

The observations of a few herpetologists (amphibian and reptile specialists) who have dedicated their careers to long-term population monitoring, dealing with many individuals, offer some answers. Table 1.1 summarizes the bulk of the existing data on natural longevity in snakes. The scarcity of data allows for few generalizations. Contrary to expectations, however, small-bodied species were not found to have shorter life spans than large-bodied species.

Longevity is a function of a snake's physiology and the harshness of the environment; both determine the probability of survival from one year to the next. Despite incomplete data, for instance, advanced old age appears to be rare among hook-nosed seasnakes because of high annual mortality. By contrast, annual mortality is much lower among terrestrial ringneck snakes, and more individuals live to older ages.

WHAT DOES IT MEAN TO BE COLD-BLOODED?

"Cold-blooded" is a colloquial way of saying *ectothermic*, a term that describes how an animal obtains its body heat. "Warm-blooded" birds and mammals are *endothermic*, meaning that their body heat is produced inside the animal by means of cellular metabolism (*endo,* inside). All other animals are *ectothermic* (*ecto,* outside), deriving their body heat from external sources such as the sun, air, water, or ground.

Two related physiological terms, *homoiothermy* and *poikilothermy*, also refer to body temperature, but to its constancy rather than its source. Homoiothermic animals, including mammals, maintain a near-constant body temperature independently of their surroundings. In contrast, the body temperatures of poikilothermic animals like snakes vary continually or occasionally with environmental temperatures.

Cold-bloodedness involves both ectothermy and poikilothermy. Although the two phenomena tend to go hand in hand, they are not always linked. A hummingbird, for example, is a poikilothermic endotherm. Like all birds, it has a high metabolism and generates its own body heat. But because it is small, it loses heat rapidly. To minimize heat loss while sleeping, it lowers its metabolism and allows its body temperature to drop, thus conserving energy.

Among snakes there are no homoiothermic ectotherms, although in some environments they might approach that condition. Many snakes do maintain a near-constant body temperature during part of each day.

Why Do Snakes Bask in the Sun?

Ectotherms have poor insulation. Insulation—fur, feathers, body fat, or oils—functions to prevent both heat loss and heat gain. Insulation would be counterproductive on an ectotherm because it must absorb heat from an outside source, and the sun is the world's primary heat source. Basking raises body temperature, and in terms of energy expenditure, it is an economical means of gaining heat. Many ectotherms bask, including tropical as well as temperate-zone species.

Some ectotherms, known as thermal conformists, allow their body temperature to follow the temperature of their surroundings. Thermal nonconformists, by contrast, maintain a constant body temperature during part of their daily or seasonal activity cycle. These adaptations have arisen over countless generations of selection for a particular physiological and behavioral lifestyle. Both lifestyles occur among snakes, but, depending on seasonal weather conditions, a nonconformist species may fall into the pattern of conformity for a short period of time, and vice versa.

Are Snakes Ever Warm-Blooded?

Lacking body insulation, snakes use behavioral and physiological mechanisms to elevate and maintain a near-constant body temperature. Each day, such a thermal-nonconformist snake will elevate its body temperature to the preferred level and maintain it within 1 to 2°C throughout its activity period. Many snakes are "warm-blooded" in this nontechnical sense. Thermal-nonconformist snakes, such as *Coluber* and *Elaphe*, have preferred body temperatures in the range of 24 to 36°C. The preferred temperature varies not only by species but seasonally within a species.

Because chemical processes speed up with rising temperature, an elevated body temperature will speed digestion. But do snakes bask to improve their digestion? The answer is uncertain; perhaps some species do. After catching prey, some species select a retreat that allows them exposure to the sun. There may be an association between species that bask (or seek to elevate body temperature) and species that eat large prey and thus require extended time for digestion.

Each morning, the thermal-nonconformist snake emerges from its resting site, moves to its basking site, orients its body to expose maximum surface to the sun's rays, and physiologically increases its cutaneous (skin) blood circulation to absorb heat and transport it to the body core. Heat is absorbed directly from the sun's radiation, from the heated stone's surface, and from the air. Having reached its preferred body temperature, the snake begins its daily foraging. By reducing cutaneous circulation and shuttling in and out of the sun as it searches for prey, it maintains its temperature within a narrow elevated range as long as conditions permit. If it is an ambush hunter, its chosen ambush site probably permits some sunlight to hit the snake; by coiling, the snake also reduces its surface area and retards heat loss. If a nonconformist snake is unable to reach its preferred temperature because of weather conditions, the snake still pursues its daily activities but less efficiently because muscle activity and other physiological processes operate less efficiently outside the snake's preferred temperature range. If its surroundings become either too cool or too hot, the snake retreats to its resting site, which usually has a stable temperature regime both day and night.

Most thermal conformists, by contrast, live in habitats where it is impossible, difficult, or unsafe to elevate and maintain a near-constant body temperature. A rain-forest-floor snake would find few patches of sun in which to bask, and the search for such patches and necessity of moving constantly as the patch moves would increase exposure to predators. Burrowing and aquatic snakes are in constant contact with soil or water, both of which rapidly drain heat from noninsulated objects such as snakes. Similarly, nocturnal snakes lack ready sources of heat to elevate their body temperatures. But conformist species do not entirely ignore temperature differences in their environments. For instance, nocturnal snakes near highways use the road surface, often at risk of their lives, to raise their body temperature. Aquatic snakes hunt in shallow water in spring and fall when it is warm and in deeper water in summer when the shallows become too hot.

HOW DO SNAKES SPEND THE WINTER?

With the arrival of colder temperatures in the temperate zones, snakes' activity levels decline. Because snakes, unlike birds, are incapable of migrating long distances to escape the hardships of winter, they must remain in the vicinity of their summer activity ranges. They survive the winter by becoming inactive and hibernating.

Although prey is scarce in winter, hibernation is probably triggered by low temperatures rather than lack of food. As ectotherms dependent on the environment as their heat source, snakes must respond to seasonal temperature fluctuations through behavioral means. Most behavior changes involve their patterns of daily activity, such as becoming more active during daylight in the spring and fall and becoming more nocturnal during the hottest period of the summer.

Winter inactivity characterizes nearly all temperate-zone snakes. By seeking a secure hibernaculum (den) where temperatures remain low but above freezing, snakes can survive long periods of low temperatures that would otherwise incapacitate and possibly kill them. Hibernation involves such physiological changes as lowering of the metabolic rate and changes in the composition of the blood, as well as behavioral changes such as cessation of feeding (aphagia) and lethargy, followed by torpidity and dormancy.

Most snakes are probably facultative hibernators whose winter inactivity is triggered by external cues such as dropping air temperatures. Some, however, appear to be obligatory hibernators, in which some endogenous (internal) control mechanism—a biological clock—initiates winter inactivity at approximately the same time each year. In Utah, for instance, striped whipsnakes (*Masticophis taeniatus*) return to their hibernacula every year a nearly constant number of days after their spring emergence. Those that emerge early return to the hibernaculum first, and those that emerge late return last. In the laboratory, European asps (*Vipera aspis*) become inactive and attempt to hibernate at about the same time each year even when allowed to select from a series of thermal regimes, including warm temperatures. In the common gartersnake (*Thamnophis sirtalis*), metabolic depression of the heart muscles occurs only in fall and winter, their normal hibernation period.

Many temperate-zone snake species will not reproduce unless they undergo a period of cooling and metabolic depression; this response may be linked to endogenous annual rhythms. Many North American and European snakes breed in the spring; their reproductive zeal seems to be associated with rising temperatures, but apparently they must first be cooled to respond to the rising temperatures of spring.

Do Snakes Hibernate Like Mammals?

Some physiologists use the term *brumation* to describe the period of winter inactivity in reptiles and other ectothermic animals. They reserve *hibernation* for endothermic birds and mammals that experience physiological changes in blood composition and volume, heart and breathing rates, and cellular metabolism. They believe that reptiles undergo no major physiological changes other than those associated with a general slowing of physiological processes. Recent research has shown, however, that many hibernating reptiles also experience major physiological changes. Thus it is correct to say that snakes hibernate.

Most temperate-zone snakes can initially compensate for falling temperatures by slightly increasing their metabolic rates, but eventually their metabolic rates decline as the temperature falls farther. Elevation of an animal's metabolic rate is too energy-demanding to sustain for long without insulation or prey in winter.

Metabolism, which is low in hibernating snakes but not turned off, is fueled mainly by glycogen, a storage form of glucose, in the liver. Liver glycogen levels are high in the fall before hibernation and slowly decrease over the winter, as does liver weight. Meanwhile blood-sugar levels increase through the conversion of liver glycogen, indicating the use of sugar as a metabolic fuel in winter. Total body lipids—the major energy source at all times—also appear to decrease during hibernation.

Snakes emerging in the spring usually exhibit weight loss, much of which is attributable to loss of body water as well as depletion of energy stores. In fact, terrestrial snakes overwintering on land are often quite dehydrated when they leave their hibernaculum in the spring.

Where Do Snakes Hibernate?

Selection of a hibernation site is critical for a snake's survival. Winter mortality can be high in some populations, especially if good hibernation sites are scarce or a winter is particularly severe. Where hibernacula are readily available, however, more than 90% of a population will survive the winter's cold. The hibernaculum must provide shelter from freezing temperatures and from predators. It also needs enough moisture to prevent a critical loss of body water (desiccation) during hibernation. Some airflow is necessary, but lowered metabolism lessens the need for oxygen and most snakes can tolerate brief periods without breathing.

Snakes hibernate either singly or in aggregations, depending on the species and the habitat. A few snakes simply burrow into loose soil during the winter (hognose snakes, *Heterodon*), but because snakes have limited digging ability, most use pre-existing cavities as hibernacula. Often they merely crawl under or inside objects such as large rocks, logs, or stumps, or among the roots of windfelled trees. Such sites are typical for small colubrid snakes. Mammal burrows, especially those of rodents, are often used, as are the burrows of crayfish (massasaugas, *Sistrurus catenatus*) or tortoises (eastern diamondback rattlesnakes, *Crotalus adamanteus*; indigo snakes, *Drymarchon corais*), and ant mounds (small colubrid snakes, such as *Opheodrys*, *Storeria*, and *Thamnophis*). Rock crevices and cracks in rock walls, building foundations, and old wells are also used by many species. Where winters are mild, arboreal species may hibernate in elevated tree holes. Migration is often necessary if a given area cannot offer both hibernation sites and summer foraging areas. Migration distances may vary from as little as a few hundred meters up to 10 to 20 kilometers.

When snakes use a communal hibernaculum, it is probably the only suitable refuge in the area. Some aggregations, such as rattlesnake (*Crotalus*) and viper (*Vipera*) dens, consist predominantly or exclusively of a single species. Other hibernacula shelter several species of snakes. For example, anthill aggregates have been found to house red-bellied snakes (*Storeria occipitomaculata*), smooth greensnakes (*Opheodrys vernalis*), and common gartersnakes (*Thamnophis sirtalis*).

Fall aggregation is usually poorly synchronized: the members of a communal den often take a month to arrive at the hibernaculum. Juveniles and neonates usually arrive at the same time as adults; they also emerge with the adults in the spring. Visible snake density around the den builds to a peak in mid-autumn and then declines as more and more individuals enter the hibernaculum and do not re-emerge. As long as the weather remains mild, however, individuals bask in the vicinity of the den, and some species might emerge briefly during a midwinter warm period.

Snakes do not simply select a location in the den and remain motionless. Whether hibernating singly or in an aggregation, their location in the hibernaculum is temperature-dependent. They move deeper, episodically, as outside temperatures decline, remaining stationary only when temperatures are near or below freezing and they become truly torpid. As outside temperatures rise, the snakes gradually shift toward the entrance. In spring, snakes commonly emerge from hibernation at temperatures lower than those that initiated hibernation in the fall.

Do Snakes Huddle Together to Keep Warm?

Only two decades ago, the consensus was that winter denning or aggregation was an adaptation to reduce heat loss and avoid freezing during the winter. Now we are less certain that an aggregation of snakes can maintain a temperature above that of the surrounding soil and rocks throughout the winter. Aggregation slows initial heat loss as hibernation begins, but eventually, despite its reduced surface area, the aggregation's temperature will match that of the surrounding den. Hibernating snakes are not immobile, but their limited movement combined with their lack of insulation appears not to produce enough heat to maintain an elevated temperature for the aggregation. In these circumstances and even those with smaller numbers of snakes, a more important function of winter aggregation is to bring females and males of some species together for mating, either before hibernation or upon emerging.

Massive hibernating aggregations have been observed in some areas (as many as 10,000 *Thamnophis sirtalis* in one hibernation site), apparently because there are few refuges deep enough to avoid subfreezing temperatures.

HOW DO SNAKES REPRODUCE?

Some snakes lay eggs; others give birth to fully formed young. The great diversity of snakes, however, blurs the line between egg-laying (*oviparity*) and live-bearing (*viviparity*). Some snakes lay eggs whose embryonic development has just begun (ratsnakes, *Elaphe*). Others lay eggs whose development is nearly complete and that will hatch within days to a few weeks (smooth greensnake, *Opheodrys vernalis*). Some snakes retain fully yolked eggs without shells in the oviducts or uteri (boas: *Boa, Epicrates*), but the developing embryos have no connections to maternal tissue. In other species that retain eggs, the embryos develop fully functional placentas with the mother (common gartersnake, *Thamnophis sirtalis*). Similar diversity characterizes all other aspects of snake reproduction.

How Do Snakes Find Their Mates?

In general, snakes rely on smell to locate their mates. Reproductive odors for tracking and courtship are sensed by the vomeronasal organs (see *Can Snakes Smell?*). These odors arise mainly from minute skin glands on the backs of snakes. Snakes also have paired cloacal glands, which produce recognizably different odors in different species, apparent even to humans (see *How Do Snakes Defend Themselves?*). The secretions of these cloacal glands probably serve a reproductive purpose as well as a defensive one. Biologists have only recently recognized the possible role of these odor compounds, called pheromones, in courtship in snakes. Pheromones probably occur in most snakes and probably stimulate courtship, but experiments have demonstrated this function in fewer than a dozen species.

Among solitary species, males' movements become more wide-ranging when they are sexually ready. If a male encounters a female's trail, the pheromone trail tells him the direction of the female's movement and whether she too is sexually ready. He follows her trail; when he finds her, he begins to court her. Courtship in most snakes involves contact; the male crawls over and beside the female, often rubbing or tapping his chin against the female's back and head. This contact increases the female's receptivity to copulation, and in some species stimulates ovulation of the eggs from the ovary into the oviducts. When the female is receptive, she lifts her tail. The male then twists his tail under her tail so that the openings (vents) of their cloacae meet.

Male lizards and snakes have two copulatory organs, the *hemipenes*, one on each side of the base of the tail, at the edge of the vent. This feature, unique to snakes

Mating common garter-snakes (*Thamnophis sirtalis*) (M. Khosravi and N. Simpson).

and lizards, reveals their evolutionary ties. When not in use, each hemipenis turns inward like a finger of a glove turned outside-in; thus the outside surface of the hemipenis is inside. When the male everts one of its hemipenes into the female's cloaca, it unfolds. The base emerges and enters the cloaca first; when the tip emerges, it is deeply positioned in the female's cloaca. Because the surface of the hemipenis is covered with folds, flounces, or spines, depending on the snake species, this unfolding process is the only possible means of insertion and withdrawal. Copulation usually lasts less than an hour but can persist for more than a day.

Many species that aggregate for hibernation mate either immediately before or immediately after hibernation. The most dramatic early-spring mating events occur at the massive hibernation sites of red-sided gartersnakes (*Thamnophis sirtalis parietalis*) in Manitoba. Hundreds of gartersnakes use the same wintering dens. In the spring, males emerge first but do not leave the hibernation site. As females emerge, usually singly, they are mobbed by courting males. Somehow one male gains a female's acceptance and copulates with her. Meanwhile the other males scatter. The successful male releases a pheromone that not only declares his success but also creates temporary impotence in other males if they linger too long in contact with the mated pair.

Rattlesnakes and many other viperid and colubrid snakes engage in male–male combat, rather a pushing and toppling contest, to determine which individual is the dominant one and gains access to a reproductively receptive female.

In many snakes (such as kingsnakes, rattlesnakes, and taipans), males compete aggressively when they simultaneously find a sexually ready female. In the resulting *combat dance*, the combatants typically hold their heads and forebodies erect while each attempts to force the other's body downward. The pushing and shoving continue for many minutes; eventually one male establishes his dominance over the other and the losing male departs. The successful male may find the female still nearby, but he is more likely to have to continue trailing her.

How Often Do Snakes Mate?

Most snake species mate and produce young once a year. Some species mate in the spring, others in late summer or fall. The female may copulate several times during a single breeding season, with the same male or different males, if she has opportunities to do so. This behavior apparently ensures that she has adequate sperm, possibly from two or more males, to fertilize her eggs and adds genetic diversity to her offspring.

Not all species produce a clutch of eggs or a litter of offspring each year. Some species cannot reproduce annually because of energy demands. Live-bearing females do not eat, or eat sparingly, during pregnancy. Instead they use their fat and other energy stores, and by the time the young are born the female is emaciated. Since birth commonly occurs in late summer or fall, the female has little time to feed before hibernation. It may take a year or more before the female has stored enough energy to again become pregnant and undergo the semistarvation of pregnancy. Many North American rattlesnakes (*Crotalus*) have breeding–birth cycles of 2 to 4 years, and in a few populations females reproduce only at 5-year intervals.

To ensure reproductive success, the females of many snake species do not mate the same year that they give birth. The females of these species can store sperm for future use in microscopic sacs in the walls of their oviducts. The sperm, which probably draw nutrients from secretions produced in the glandular sac walls, stay viable for several years. For example, a captive indigo snake (*Drymarchon corais*) laid fertile eggs 4 years after her last mating, and a banded cat-eyed snake (*Leptodeira annulata*) did so after 6 years. Most temperate-zone snakes store sperm for several months, because copulation takes place before the eggs mature.

One species of snakes even reproduces without mating. The brahminy blindsnake (*Ramphotyphlops braminus*) is an all-female species. Its eggs are self-fertilizing and produce only female offspring, identical genetic replicas of their mother. This process is clonal reproduction, or *parthenogenesis*.

Do Female Snakes Brood Their Eggs?

Only a few species of snakes stay with their eggs. Current evidence indicates that only the carpet python (*Morelia spilota*), Indian python (*Python molurus*), water python (*Liasis fuscus*), and perhaps some other python species truly brood their eggs like birds. Carpet, water, and Indian pythons generate heat by means of minute muscular contractions, somewhat akin to mammalian shivering.

Brooding to maintain elevated incubation temperatures requires behavioral as well as physiological adaptations. A female chooses a secluded nest site buffered from daily temperature fluctuation. The entire clutch of eggs is laid at one time, and the female immediately coils around them. Brooding temperatures are 30 to 34°C in the Indian python and 28 to 33°C in the carpet python. "Shivering" maintains the elevated incubation temperature when the ambient temperature declines. At least in the Indian python, normal embryonic development requires an elevated incubation temperature; abnormal development results when eggs incubate at lower temperatures.

How Many Eggs or Young Are Produced?

Clutch sizes range from a single egg to more than 100 eggs, and litter sizes range from 1 to more than 150 newborns. The number of eggs or young varies by species; not surprisingly, small species produce fewer eggs than large species, and slender

A female blood python (*Python brongersmai*) broods her eggs. Her coils serve to insulate the eggs and maintain fairly constant incubation temperatures. (David G. Barker)

species produce fewer than heavy-bodied species. There are exceptions, however. For example, ringneck snakes (*Diadophis punctatus*) and brownsnakes (*Storeria dekayi*) are nearly the same size in adulthood (20 to 30 centimeters total length) and live in the same forest, yet ringneck snakes average fewer offspring (3 to 4 eggs) than the slightly smaller brownsnakes (14 to 15 fetuses). Live-bearers tend to have more offspring than egg-layers, but there are exceptions.

Heredity determines the upper limit of how many ovarian eggs can develop in each species. The maximum possible number occurs only rarely, because egg production is energetically expensive. Ultimately, the number of eggs depends on a female's prey resources and her health. Maximum clutch or litter sizes are seen in healthy, well-fed zoo snakes, but rarely in wild ones. Occasionally someone will find a massive clutch of 20 to 50 eggs of a species that seldom lays more than a dozen eggs, such as an American ratsnake (*Elaphe obsoleta*) or American racer (*Coluber constrictor*). These clutches are communal nesting sites in which several individuals lay eggs in the same cavity or hole.

Available prey resources, and the ability to harvest them, also determine the frequency of offspring production. If resources are low, or if a female is injured or diseased, she will have insufficient fat stores to produce eggs and will skip an egg-laying season. If resources are high, some species are genetically capable of producing more than one clutch of eggs in a nesting season. Egg-layers regularly produce at least one clutch of eggs each season; many live-bearers are on a 2- to 3-year cycle, although some (such as *Nerodia* and *Thamnophis*) produce young each year.

In an egg-laying female, the eggs mature quickly (4 to 6 weeks) and are laid soon thereafter, permitting the female to begin foraging (Table 1.2). By contrast, the burden of pregnancy prevents the live-bearing female from eating much if at all for 2 to 3 months before the birthing (e.g., timber rattlesnake, *Crotalus horridus*). Replacing her depleted energy stores in order to repeat the reproductive process may take a year or more.

DO FEMALE SNAKES PROTECT THEIR YOUNG?

The oft-told tale of a female snake swallowing her young to protect them has never been verified. Thousands upon thousands of snakes have been kept in captivity and such behavior has never been observed. Similarly, thousands of dissections for dietary studies have never revealed undigested young in a female's stomach.

Some snake species do, however, exhibit various degrees of parental care. Attendance of the eggs or newborn, the most basic form of parental care, is a widely distributed but not commonplace behavior throughout the various snake groups. In

TABLE 1.2. DURATION OF INCUBATION OR PREGNANCY IN SELECT SPECIES OF SNAKES

Duration of incubation or pregnancy (days)	Species	Family or subfamily	Where recorded
Egg-laying (oviparous) snakes			
3–5	Himehabu (*Ovophis okinavensis*)	Crotalinae	Okinawa
14–30	Smooth greensnake (*Opheodrys vernalis*)	Colubrinae	Virginia
22–47	Rough greensnake (*Opheodrys aestivus*)	Colubrinae	Virginia
28–56	Grass snake (*Natrix natrix*)	Natricinae	Austria
30–45	Schlegel's blindsnake (*Rhinotyphlops schlegelii*)	Typhlopidae	Zimbabwe
45–46	Eastern wormsnake (*Carphophis amoenus*)	Colubrinae	Virginia
54–68	Eastern racer (*Coluber constrictor*)	Colubrinae	Virginia
55–67	Checkered keelback (*Xenochrophis piscator*)	Natricinae	India
57–61	Water python (*Liasis fuscus*)	Pythonidae	Queensland
57–66	Indian python (*Python molurus*)	Pythonidae	Sri Lanka
60–80	Common taipan (*Oxyuranus scutellatus*)	Hydrophiinae	Queensland
90–97	Ball python (*Python regius*)	Pythonidae	Uganda
105–108	Australian scrub python (*Morelia amethystina*)	Pythonidae	Queensland
Live-bearing (viviparous) snakes			
70–84	Common nightadder (*Causus rhombeatus*)	Viperinae	Ghana
70–85	Rough earthsnake (*Virginia striatula*)	Natricinae	Virginia
100–140	American copperhead (*Agkistrodon contortrix*)	Crotalinae	Kansas
105–113	Dekay's brownsnake (*Storeria dekayi*)	Natricinae	E. North America
110–350	Common European viper (*Vipera berus*)	Viperinae	Great Britain
120–180	Jararaca (*Bothrops jararaca*)	Crotalinae	Brazil
152–252	Cuban boa (*Epicrates angulifer*)	Boinae	Cuba
180–240	Terciopelo (*Bothrops asper*)	Crotalinae	Costa Rica

Birth in a Caucasian viper (*Vipera kaznakovi*). The neonate has just been expelled from the female's cloaca and is still encased in its amnion sheath.

all instances that have been examined carefully, the attending adult was a female. Attendance of eggs has been reported for one or two blindsnakes, various colubrids (*Amphiesma, Elaphe, Farancia*), and a few vipers (*Calloselasma, Trimeresurus*) and elapids (*Micrurus*). In some cases, the attending female displays aggressive behavior in defense of her egg clutch. King cobras (*Ophiophagus*) and common cobras (*Naja*) appear to defend a small territory around the eggs; others, such as kraits (*Bungarus*), defend only when the nest and resident female are uncovered. The ultimate in parental care occurs in the pythons, some of whom brood their eggs and can generate heat to maintain a high-incubation temperature (see *Do Female Snakes Brood Their Eggs?*).

Among live-bearing snakes, female attendance of newborns is more ambiguous because it is difficult to distinguish protection from female presence owing to the recency of birth and slow dispersal of the newborn. In some pitvipers and true vipers, there appears to be a delay of several hours or more before the female and newborn disperse. Even such limited attendance conveys some protection and can be considered parental care.

DO SNAKES HAVE ENEMIES?

Snakes have many natural enemies. They prey on many vertebrates and invertebrates, and are in turn prey for most predatory vertebrates and many invertebrates as well.

Spiders are probably the main invertebrate predators of snakes; they habitually catch and eat small juveniles and small-bodied species. In Brazil, the large *Grammostola* spider has sufficient venom to kill even small pitvipers. In the United States, small snake species (such as ringneck, brown, and lined snakes) are occasional prey of spiders. Scorpions are snake predators in deserts, and even some large insects, such as aquatic giant waterbugs (Hemiptera), attack small snakes. Some ants, particularly the fire ant introduced into the southeastern United States, attack snakes; they are particularly destructive to the eggs of oviparous species.

Few predatory fish will pass up a swimming snake of manageable size. Various frogs, toads, turtles, lizards, and crocodilians also eat snakes. But of all ectotherms, snakes are the prime predators of other snakes. Indeed, the elongated limbless body of a serpent makes it ideal food for another serpent. Snakes usually eat smaller snakes, but occasionally a snake will eat another snake larger than itself. In such cases the ingested snake is folded over on itself within the digestive tract of the predator. The larger the snake, the larger the snake it can eat: a 2.9-meter black mamba (*Dendroaspis polylepis*) was seen eating a 2.3-meter cobra (*Naja melanoleuca*), and a 3.9-meter king cobra (*Ophiophagus hannah*) was observed swallowing a 2.4-meter python.

Some species of snakes are almost exclusively snake-eaters (ophiophagous). Many cobras, kraits, and coralsnakes are ophiophagous, particularly the king cobra. Many colubrid snakes are also ophiophagous, and some, such as the North American kingsnakes (*Lampropeltis*) and the tropical American mussurana (*Clelia*), are immune to pitviper venom and regularly prey on pitvipers.

Some snakes do not prey on other snakes but do eat snake, lizard, and turtle eggs. The scarlet snake (*Cemophora coccinea*) is one such egg-eating (oophagous) specialist.

Birds are prominent snake predators. Most predatory birds include snakes in their diet. A few, such as the American red-shouldered hawk and the European short-toed eagle, seem to intensify snake-collecting during the nesting season to feed their nestlings. A few birds specialize on snakes and lizards. Probably the best known is the African secretary bird, which walks through savanna and scrub. When it comes on a snake, the secretary bird pummels it to death with its feet, occasionally tossing it in the air with its bill. The roadrunner of the American Southwest similarly attacks and eats snakes.

The skeleton of a rattlesnake attests to the adage that every species, even predators, is the prey of another species.

Predatory mammals seldom turn down a meal of snake. But of all animals, the most destructive to snakes are humans. We are chiefly responsible for the dwindling populations of snakes throughout the world. We prey directly on snakes for food, skins, and medicinal products. Since the beginning of recorded time, humans have practiced predator control (often by means of bounties) on snakes. Habitat modification—whether for agriculture, forestry, commerce, or housing—has made many areas unfit for snakes. Automobiles also kill large numbers of snakes and other wildlife along major roads. Humans have been and remain a serious threat to the survival of snakes in most parts of the world.

Finally, natural catastrophes, such as grassland and forest fires, avalanches, volcanic eruptions, floods, extreme droughts, and sudden extreme drops in temperature take their toll on snakes.

WHAT SOUNDS DO SNAKES MAKE?

Snakes have a voicebox (larynx), and some may have vocal cords, but they cannot vocalize. Nevertheless, snakes are not mute or silent, for they produce a variety of sounds, mainly while defending themselves. The most familiar of these sounds is the hiss, and it may be that all snakes hiss. Certainly the larger species do, but if small species such as the blindsnake hiss the sound is not audible to the human ear.

A hiss is produced by filling the lungs and then rapidly constricting the body wall to force the air out through the glottis, usually with the mouth open. A hiss can sound like a slowly leaking punctured tire or it can be much louder, like the rapid escape of steam from heating pipes. The large African vipers, such as the puffadder (*Bitis arietans*) and the Gaboon viper (*B. gabonica*), are well known for their "steam-heat" hisses. In these vipers and other snakes, the sound arises from air forced through the labial notch or other narrow openings. In the North American bull-snake (*Pituophis*), the air exits forcibly through the glottis and a partially open mouth. Hissing in other species of snakes has been said to resemble a baby crying, a cat meowing, a dog growling (king cobra, *Ophiophagus hannah*), and even a musician's tuning fork. The king cobra's growl is much lower-pitched than the typical snake hiss and is probably produced by a special resonating chamber in the cobra's tracheal lung. In contrast, the "startling hiss" of the timber rattlesnake (*Crotalus horridus*) is a sudden inhalation of air, alerting the intruder to the snake's presence while inflating and enlarging the snake.

How Does a Rattlesnake Rattle?

Aside from a hiss, the best-known snake sound is the rattling of American rattlesnakes. Only species of *Crotalus* and *Sistrurus* produce the rattling sound. Rattling seems a misnomer for the high-pitched buzz of a rattlesnake. In most rattlesnakes, the rattle has a peak or dominant frequency of about 9,000 hertz (ranging from 2,500 to 19,000 hertz), roughly equivalent to that of an ambulance siren, and a loudness of 60 to 80 decibels at a distance of 1 meter, like that of a baby's rattle. Both pitch and loudness signal danger to humans and other animals. Rattling is a warning to stay away or be bitten.

The rattling sound arises from rapid vibration of a series of loose-fitting, interlocking, cone-shaped scales on the tip of the tail. Vibration of the tail at about 50 cycles per second causes the edges of these scales to rub against one another, producing the spine-chilling rattling. Pitch and loudness differ between species and among individual snakes. Larger individuals and species produce louder, lower-pitched rattling. Snakes' body temperatures affect the rate of vibration: the rattling of a cold snake is quieter and lower-pitched; loudness and pitch increase as a snake approaches its preferred body temperature.

The tail and rattle are held erect and vibrated to give warning if this speckled rattlesnake's (*Crotalus mitchellii*) space is further invaded.

At birth, a rattlesnake has a single rattle segment, the prebutton. The prebutton is lost with the snake's first skin molt, usually within 2 weeks of birth, and is replaced by the underlying *button*, which becomes the first retained rattle. Thereafter, unless it is broken off, the button is the end of the rattle. Each time the snake sheds its skin, another rattle segment is added to the base of the rattle. If the button is intact, the number of rattle segments plus the button indicates the number of times the rattlesnake has shed, but not the age of the snake; a snake can and often does shed more than once a year (see *Why and How Do Snakes Shed Their Skin?*).

No one knows how and why the rattle evolved. There has been much speculation on the evolution of the rattle, but one hypothesis now prevails. Many snakes, not just rattlesnakes, nervously shake their tails when frightened or threatened. If the snake is in dry grass or leaves, the tail vibration produces a sound similar to the rattlesnake's buzz. This sound alerts the predator or interloper that the snake is alert and ready to defend itself. The ancestors of rattlesnakes presumably evolved on the grass-covered plains of North America, populated by herds of large herbivores. A snake in the grass is hard for herbivores to see, but one rattling in the grass announces its presence. If its venomous bite causes discomfort, herbivores will learn to avoid stepping on a rattling snake. This hypothesis suggests that, over time, the rattle evolved by selecting for the individuals with better sound-making capabilities.

Do Snakes Make Other Sounds?

Several snakes make sounds by rubbing their body scales together. The best known scale-rubbers are the saw-scaled vipers (*Echis*), but some horned vipers (*Cerastes*) and the African egg-eating snakes (*Dasypeltis*) also do so. All scale-rubbing snakes live in deserts or dry grasslands, and scale-rubbing might also have evolved as a warning device to protect snakes from being trampled by large herbivores. These snakes have several rows of heavily keeled (ridged) scales oriented obliquely on each side of the body. The keel extends the full length of the scale and is often serrated (saw-toothed) along its upper edge. When aroused, the snake inflates its body and adopts a figure-eight posture. As the snake moves into this posture, the keeled scales on the surface of one loop rub against those on an adjacent loop, producing a loud rasping sound. The snake repeatedly coils and uncoils, making a nearly continuous rasping that can be quite intimidating—especially if you suspect the noise to emanate from a saw-scaled viper, one of the world's most deadly snakes.

The strangest sound made by a snake is the cloacal "pop" of the small North American western hooknose snake (*Gyalopion canum*) and the Sonoran coralsnake (*Micruroides euryxanthus*). When disturbed, the hooknose snake begins writhing; in its wild gyrations it can even throw itself up in the air. While writhing, the snake continually extrudes and retracts the lining of its cloaca through the vent, alternatingly sucking in and blowing out air. This air movement produces a bubbling or popping sound.

HOW DO SNAKES DEFEND THEMSELVES?

"Snakes are first cowards, next bluffers, and last of all warriors" (Pope 1958). This statement aptly summarizes snakes' defensive behavior. If given an opportunity to escape, a snake faced with a potential threat will quickly crawl away or take refuge in a secure hiding place. This discretion has saved the lives of many snakes. If prevented from escaping, some snakes resort to startling displays of bluff or temper. Some behaviors merely intimidate the disturber, whereas others are a clear warning of danger.

Bluffing behavior takes several forms, most of which make the snake appear larger or more aggressive than it really is. Appearing larger is often accomplished by inflating the lung to enlarge body girth, such as in hognose snakes (*Heterodon*) and some vipers (*Bitis*), or to enlarge just the throat, as in the boomslang (*Dispholidus*),

There are many ways to mislead a predator. The rough-scaled bushviper (*Atheris squamigera* [left]) bluffs with a "puffed" head and throat. The rubber boa (*Charina bottae* [right]) hides its head and waves its tail.

African birdsnakes (*Thelotornis*), red-tailed racer (*Gonyosoma*), and puffingsnake (*Pseustes*). Many species merely flatten the ribs to make their bodies appear broader (*Natrix, Nerodia*). The cobra's hood is a warning device, created by spreading the cervical ribs to flatten the throat and forepart of the body. The back of the hood of some Asian cobras (*Naja*) is adorned with one or two dark-bordered ovals resembling eyes, presumably to startle a predator and keep it from attacking the vulnerable dorsal surface of the snake's neck.

Some snakes, including the venomous cottonmouth (*Agkistrodon piscivorus*) and American vinesnakes (*Oxybelis*), display the internal lining of the mouth in a threatening gape. Humans recognize the cottonmouth display of its whitish to cream-colored oral lining—the source of its common name—as a warning that the snake is both feisty and dangerous. The red-bellied snake (*Storeria occipitomaculata*) curls its upper lip to expose its maxillary teeth while pushing the teeth outward in a toothy grin. A few snakes ooze blood (autohemorrhage) from the mouth, and the small West Indian groundboa (*Tropidophis*) squirts or drips blood from its eyes when disturbed. Some cobras spit venom as a defensive action (see *Do Snakes Spit Their Venom?*), but no venom is involved in the autohemorrhaging of the other snakes.

Several behaviors divert attention from the critical head and neck to a less vulnerable part of the body. Ringneck snakes (*Diadophis*) display a fake head by rolling the tail into a tight spiral and exposing the bright-red or orangeish-yellow ventral side while hiding their heads beneath their body coils. Other snakes simply hide the head and sway the tail from side to side. Some species hide the head among their coils, or coil into a ball with the head positioned inside (ball python, *Python*

Playing dead is a peculiar way to mislead a predator, but that is one of the eastern hognose snake's (*Heterodon platirhinos*) defensive behaviors.

regius; Calabar ground python, *Charina reinhardti*; rubber boa, *Charina bottae*; Australian graysnake, *Hemiaspis damelii*). Many head-hiders have blunt heads and tails, making it difficult to distinguish the head from the tail. This similarity may create the appearance of two heads, which would be twice as dangerous and thus likely to discourage a predator's attack.

Playing dead (*letisimulation* or *thanatosis*) presumably works by dampening the predator's interest. The best-known letisimulators are the hognose snakes (*Heterodon*). When they cannot escape, they perform a stereotyped defensive routine. First the snake spreads its neck and hisses loudly, occasionally striking with mouth closed. If this maneuver does not discourage the attacker, the snake contorts its body and writhes about, vomits up food, and finally rolls over on its back and plays dead with its tongue extended and blood often seeping from its mouth. There is, however, a flaw in its behavior: if turned over, the "dead" snake immediately rolls onto its back again. Some *Natrix*, *Storeria*, and *Virginia* also play dead, but the oddest posture is probably that of the Asian tentacled snake (*Erpeton tentaculatum*), which holds its body as straight and rigid as a board.

If picked up, most snakes release a foul-smelling mixture of musk from their cloacal glands and feces. People who have been "musked" on their hands and arms find the behavior repulsive. A predator that gets a face and mouth full of this smelly mixture will thereafter associate the sight and smell of the snake with this unpleasant experience.

In a few snakes, the scale at the tip of the tail is highly cornified and pointed like a thorn. Some snakes with spine-tipped tails are small (e.g., blindsnakes and wormsnakes); others, such as the mudsnake (*Farancia abacura*), are much larger, but all are inoffensive and rarely bite. Instead, they push the tail spine into the body of the adversary. The sharp-keeled body scales of some snakes, such as the Laurent's mountain bushviper (*Atheris hispida*), saw-scaled vipers (*Echis*), and brown watersnake (*Nerodia taxispilota*), may prevent them from being swallowed, particularly once these ridged scales are fully developed.

DO SNAKES GET SICK?

Snakes regularly get sick and die in captivity, even in the most carefully maintained captive conditions. Many of these deaths arise from secondary bacterial and viral infections common among captive animals (see "maladaptation syndrome" in *Should I Keep a Snake as a Pet?*).

Do wild snakes also experience bacterial and viral infections? Obviously the answer is yes, but our knowledge of diseases in wild snakes is limited. Observations of sick snakes in the wild are rare. Either illness impairs a snake's behavior and performance so that it is soon captured by a predator or the snake seeks shelter and remains inactive until it recovers or dies. Because we know very little about the health of snakes in the wild, much of what follows in this section applies only to captive snakes or represents extrapolation from captive snakes to free-living snakes.

What Are Some Common Diseases of Snakes?

Mouthrot or infectious stomatitis is a common affliction of captive animals. It is a bacterial infection (either *Aeromonas* or *Pseudomonas*) of a slow-healing injury of oral tissue. The initial injury might be minor (such as a tooth dislodged while eating), but if the animal's health is poor, the wound is slow to heal and a secondary infection occurs. Mouthrot arises from bacteria commonly present in the mouth and digestive tract of snakes. The infection hinders or prevents feeding, aggravating the poor health that gave rise to the secondary infection.

Pneumonia and septicemia also originate from *Aeromonas* or *Pseudomonas* bacteria. Both diseases are usually fatal, even in captivity, because recognizable symptoms appear only when the disease is well advanced. These and other bacteria can cause cell damage in organs such as the kidneys, liver, skin, and digestive tract. In most instances, infection follows declining health. Natural disasters, such as fire and drought, can stress wild snakes and make them susceptible to these diseases. Fungal (mycotic) infections cause similar diseases, and also seem to be secondary infections owing to poor health. Tuberculosis in snakes is caused by a mycobacterium.

Viral diseases are less studied in snakes than bacterial and fungal illnesses. Nevertheless, viral infections are likely to be as prevalent in snakes as they are in birds or mammals, and some evidence suggests that snakes are susceptible to mosquito-transmitted viral encephalitis.

Do Snakes Have Parasites?

External parasites, especially mites and ticks, are often seen on wild snakes. Mites are tiny blood-sucking arthropods. Most are integumentary parasites that attach themselves to the soft skin between the scales. A few mites will be found on most wild snakes, but heavy infestations are rare. In captive snakes, most mite infestations arise from improper quarantine, failure to disinfect new snakes, and poor maintenance of the captive environment. Since a captive snake is confined, it cannot escape constant reinfection, and infestation tends to increase and increase. Eventually the number of mites on the snake becomes life-threatening owing to blood loss. At least one type of mite lives in the trachea and lungs of snakes. Chigger-mites, which burrow superficially beneath the skin, are as irritating to snakes as they are to humans.

Ticks are larger blood-sucking arthropods that also attack wild snakes. Snakes do not appear to be the primary host of any species of tick; instead they are intermediate hosts for one or more stages of mammalian ticks.

Both ticks and mites cause discomfort, but they and mosquitos—yes, snakes are bitten by mosquitos and other blood-sucking insects!—can also transmit viral and protozoan diseases to snakes. Numerous blood protozoans have been found in snakes; many are transmitted from host to host by biting insects. These blood parasites include flagellates, such as *Trypanosoma* and *Leishmania*, and coccideans. The coccideans include the malaria-causing *Plasmodium* parasites, which are widespread in tropical and subtropical lizards and probably in snakes as well. The effects of other blood protozoans on health are largely unknown.

Specialized blood-feeding arthropods called pentastomids or tongue worms have also been found in the mouth, pharynx, and lungs of rattlesnakes (*Crotalus*), the cottonmouth (*Agkistrodon piscivorus*), and some pythons.

A huge assortment of intestinal protozoans have been found in snakes. Amoebas are particularly common in the guts of snakes, as are various types of coccideans. These intestinal protozoans are likely to be harmless in healthy snakes, becoming harmful only when a snake's health declines.

Snakes also harbor an enormous variety of parasitic worms. Roundworms (nematodes) are the most common; in fact, it is nearly impossible to dissect a snake

Scales cannot be penetrated by the mouthparts of pest arthropods, but ticks and other pests attach themselves on the softer interscalar skin. (*Boa constrictor.*)

without finding intestinal nematodes. Nematodes occur in various visceral organs and muscle, but the most damaging nematodes are filarial worms that reside in the circulatory and lymphatic systems. These nematodes can become so abundant that they block blood and lymph flow. Snakes have other parasitic worms as well, including tapeworms in their guts and flukes (trematodes) in their lungs, blood vessels, and digestive and urinary tracts. Some aquatic snakes may also be harassed by bloodsucking annelids (leeches).

Do Snakes Get Cancer?

Both wild and captive snakes occasionally develop tumors. Tumors can arise from viral infections, injury, or unknown causes. Benign (noninvasive) tumors can eventually kill by blocking the digestive tract or the circulatory system, by impairing sight or locomotion, or by affecting normal function of the body in some other way. Even the slightest impairment in a wild snake increases the probability of predation.

Snakes occasionally develop pseudotumors around a parasite or bacterial infection. This sort of tumor is usually small and persists throughout the individual's life without further growth. Malignant or cancerous tumors invade adjacent tissue and kill or displace the normal cells, impairing their functions. Eventually the cancerous tumor metastasizes (spreads), sending cancerous cells throughout the body where they establish new tumors. A variety of cancers has been reported in snakes, including leukemia, melanomas, and adenocarcinomas.

Do Snakes Feel Pain?

All vertebrates possess nerve endings and other receptors that register pain in the skin, muscles, and elsewhere. We can thus be certain that snakes feel pain—perhaps with the same intensity as do humans and other animals.

Many humans allow severe pain to incapacitate them, but few animals do. All vertebrates are subject to shock, which is widespread failure of the body's physiological mechanisms due to severe cell, tissue, and/or organ damage. We do not know, however, whether nonhuman animals can be psychologically incapacitated by pain.

WHEN DID THE FIRST SNAKE APPEAR?

Snakes appeared in the Cretaceous, more than 120 million years ago. Many features of snake anatomy suggest that they arose from a lizard ancestor via an evolutionary stage that involved burrowing. Limblessness, loss of external ears, and loss of color receptors (cones) in the retina are a few traits suggesting an early burrowing stage. However, an alternate hypothesis proposes the initial lizard-to-snake stage as an aquatic one. Only in the mid-1990s was a fossil proposed to support this transition. *Pachyrhachis* was a peculiar long-bodied and tiny-limbed reptile. It was initially proposed as a mosasaur, but upon re-study, some scientists identify it as a limbed snake and possibly the ancestor to all other snakes. Other scientists disagree. Perhaps it is a snake, but it is not the ancestor to all other snakes, because snake fossils occur even earlier in the Cretaceous than it does.

The oldest known snake fossil may be two vertebrae from an Early Cretaceous (Barremian) deposit in Spain, more than 120 million years old; however, the dating of these vertebrae is uncertain. Late Cretaceous (about 96 to 100 million years ago) deposits in Africa and Spain have yielded a variety of snakes, representing at least five families. These fossil snakes include members of the extinct families Lapparentophiidae, Madtsoiidae, Nigerophiidae, Paleophiidae, and possibly one or more early colubroids. These fossil assemblages show that snake evolution was well advanced at that time. An unquestionable radiometric date of 98±1 million years applies to a Utah snake, *Coniophis*.

Another Late Cretaceous snake fossil, *Lapparentophis defrenni*, found in the Sahara Desert of Algeria, consists of just three vertebrae that share many features with those of lizards. *Lapparentophis* is thus considered a unique and primitive family (Lapparentophiidae). Unfortunately, the vertebrae offer no clues to the evolution of snakes from lizards. The Lapparentophiidae and the related fossil *Simoliophis* from the Late Cretaceous period of Europe and Egypt do not seem to be related to any living family of snakes. They apparently left no descendants. *Simoliophis*, the only

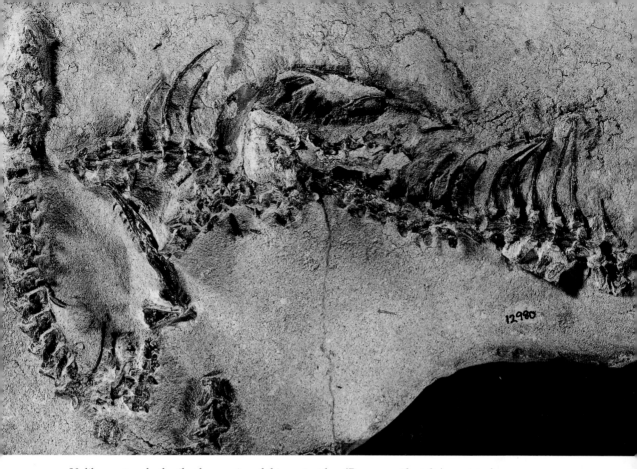

Unlike most snake fossils, the remains of this ancient boa (*Boavus occidentalis*) consist of articulated vertebrae. Even a lower jaw (center left) remained with the vertebrae. (Smithsonian Photoservices)

representative of the Simoliophiidae, might have been a marine descendant of lapparentophiid snakes. They were not, however, relatives of *Pachyrhachis*.

Two modern groups of snakes, the boas and the false coralsnakes, first appeared in the Late Cretaceous. The oldest known "boa," *Madtsoia madagascariensis*, found in Madagascar, was comparable in length to present-day large pythons and boas, although likely not a close relative of these two groups of snakes. The earliest false coralsnake (Aniliidae), *Coniophis*, is from the North American Late Cretaceous; again the identity of this fossil is uncertain and it might be a boa. Another large Late Cretaceous snake, *Dinilysia* (Dinilysiidae), was recognized from a nearly complete cranium and several vertebrae found in southern South America. The wide distribution of these early snakes suggests that several groups of snakes evolved in different areas of the world after the initial origin of snakes, a pattern that has continued to the present.

The natural history of North American snakes also begins in the Late Cretaceous with several fossil aniliids from Canada and the United States, including some from Wyoming and New Mexico assigned to the genus *Coniophis*. A possible boa (family Boidae) also dates from the Late Cretaceous of North America. After this point, the fossil history of snakes becomes much more diverse and complicated. The first caenophidians do not appear until the Eocene, about 50 million years ago. The first elapids and viperids appear in the early to mid Miocene.

Not surprisingly, vertebrae are the most common fossil remains of snakes. Because snake skeletons are delicate and held together mainly by ligaments, tendons, and muscles, snake fossils are seldom even partially articulated skeletons; more often they are isolated and widely scattered bones. The general absence of skull bones makes the identification of snake fossils difficult. Paleontologists who study fossil snakes must be able to recognize small variations in the many bones of the snake skeleton to identify accurately where the bones lie in a skeleton as well as to match the bones to a particular species. Although useful for identification of species, vertebrae are not particularly helpful in establishing evolutionary relationships, because we know from modern snakes that vertebral structure may be more closely linked to a snake's mode of life than its ancestry or relations.

HOW ARE THE AGES OF FOSSIL SNAKES DETERMINED?

To estimate the age of a fossil, it is essential to know exactly where it was found. The position of a fossil in a geologic formation, and its association with fossils of other organisms, can provide an accurate relative age, but not a precise age. Absolute age determination is possible by radiometric dating.

Radiometric methods are based on the principle that radioactive isotopes of elements slowly change (decay) to a more stable daughter element, such as lead. Because this change occurs at a constant and unique rate for each radioactive isotope, it can be used to date materials containing the isotope. As time passes, the amount of the parent isotope decreases while that of the daughter element increases. The ratio of daughter element to parent element can thus be used to determine how much time has elapsed since decay began. For any rock or mineral containing radioactive material, the starting point (time zero) is the moment when the radioactive parent atoms became part of the rock or mineral in which the daughter element is now trapped. A highly sophisticated technique, mass spectrometry, measures the proportions of minute radioactive particles according to their differences in mass (weight). Mass spectrometry is particularly useful when dealing with very old rocks; other instruments are used if the fossil and its substrate are younger. Another frequently used technique is radiocarbon dating, which depends on the accumulation of carbon-14, the radioactive isotope of carbon. Because all living organisms contain carbon, this method is a convenient way to date animal remains; however, its use is limited to materials 70,000 years or younger.

HOW MANY KINDS OF SNAKES EXIST TODAY?

There are currently more than 2,800 known species of snakes, and several new species are discovered each year. If we count subspecies, there are more than 3,000 different kinds of snakes.

How Are Snakes Classified?

The system that biologists use to classify species is hierarchical. It divides the animal kingdom into classes, within classes into orders, and within orders into families. Families are in turn subdivided into increasingly precise categories, from genus to species. Just as a genus contains one or more species, a family contains one or more genera (plural of *genus*) and all of their species, an order then contains one or more families, and so on upward through the categories of classification. Biologists now try to ensure that each category or *taxon* represents a group that shares a common ancestor. But taxa do not always meet that criterion. Generally speaking, you can assume that all species with the same generic name are more closely related to one another than any of them is to a species in another genus. This assumption applies to the higher categories as well. Even common names, such as rattlesnakes, cobras, or seasnakes, indicate real evolutionary relationships. However, common names and sometimes scientific classifications do not always correctly identify relationships. For instance, not all watersnakes (*Natrix, Nerodia*) or ratsnakes (*Elaphe, Ptyas*), or colubrids are closely related.

The scientific names we use in this book are species names. Starting in the middle of the eighteenth century, biologists recognized the need for a unique name for each species if they were to discuss their discoveries and avoid confusion. The two-part scientific name (binomial nomenclature [*bi*, two; *nomen*, name]) was born then and became the standard. A species name is always italicized, and is usually in either Latin or Greek. The first name is capitalized and the second is lowercase. The second part of the name is used for a single species of plant or animal. The first part is the generic name or genus; a hundred or more species can share the same generic name. The genus originally denoted a group of organisms that looked alike; now it and the higher classification categories denote an evolutionary relationship. Because it represents a distinct group of species, the generic name can be used without a specific name to refer to all species of that group. The specific name is never used alone; it must be accompanied by the genus name (although the genus is occasionally abbreviated to a single letter, such as *E. obsoleta*). Three-part names are

Often widely occurring snake species have different coloration and/or patterns in different geographic areas. The eastern ratsnake (*Elaphe obsoleta*) has five geographic races; two of them are the gray ratsnake (*E. o spiloides* [left]) and the eastern black ratsnake (*E. o. obsoleta* [right]).

reserved for subspecies (variants of species found in certain geographic areas), and can also be abbreviated, such as *E. o. obsoleta*.

There are more than a dozen definitions of a species. Life is diverse, and evolution and the various modes of reproduction have produced different types of species. Furthermore, different groups of biologists may view the same phenomenon from different perspectives, resulting in a diversity of definitions. Most biologists agree, however, that a species is a group of organisms with similar appearance and a unique set of genes, arising in the past from the same ancestral population. Most snake species consist of populations that are capable of producing viable and fertile offspring with one another (but not with members of a different snake species). An exception is the brahminy blindsnake (*Ramphotyphlops braminus*), which has no males (see *How Often Do Snakes Mate?*). These populations collectively constitute the species. A species is not just an abstraction but a real entity because it consists of a group of individuals related through the genetic ties of ancestry.

Some species have broad geographic distributions. In some of these species, individuals either look identical throughout the geographic range or they differ within one population in the same way they do in another population hundreds of kilometers away. The eastern hognose snake (*Heterodon platirhinos*) of North America occurs from New England to Florida and as far west as the plains of Kansas. There is no discernible anatomical difference between individual snakes from New Jersey and those from Florida or Kansas. The eastern hognose snake exhibits several color variations, but there is no geographic pattern to this variation. Thus this species lacks geographic differentiation.

In contrast, the western rattlesnake (*Crotalus viridis*) has a wide distribution in western North America and has a variety of distinctly colored and patterned populations within this area. These patterns reflect regional adaptation and geographic differentiation. Where these geographically distinct populations (that is, subspecies) meet, individuals from the adjacent populations interbreed, and the offspring are fertile and typically share a combination of the parental colorations. Such interbreeding is not hybridization because the individuals of the two populations are genetically very similar, and combination of eggs and sperm does not cause any mismatches in chromosomal and cellular division that result in failed fertilization, abnormal embryonic cell division and embryo death, or infertile offspring that survive to adulthood. All western rattlesnakes belong to the same species, but because of the different patterns among the populations, the populations are recognized formally as subspecies and are given unique names, for example, *Crotalus viridis viridis*, *Crotalus viridis helleri*, *Crotalus viridis oreganus*, and others.

How Are Snakes Related?

The classification hierarchy in Appendix 1 portrays in a rough manner the evolutionary history of snakes and the relationships among snakes. Many of the lineages shown in Appendix 1 arose early, either in the Cretaceous or the earliest Tertiary. Because of this early divergence of lineages, and the initial anatomical streamlining associated with limblessness, little distinctive physical evidence exists with which to estimate evolutionary relationships. Thus, although we have made progress, the path to understanding the relationships among snakes has been fraught with twists and reversals. Over the past two decades, a more rigorous analytical protocol and the development of molecular systematics (analysis of DNA characters) have expanded our knowledge of evolutionary relationships.

We are now reasonably certain that all modern snakes belong to two major lineages (Figure 1.4). One lineage is the blindsnakes, classified as scolecophidians (see Appendix 1). The blindsnakes are highly specialized burrowers, and most species eat termites and ants. They have been moderately successful residents of the tropics and subtropics, with more than 300 species scattered over the world, but blindsnakes have not diverged into a multitude of body forms and lifestyles.

The second lineage, the alethinophidians, is more successful, consisting of more than 2,300 species and a multitude of lifestyles. Even though some alethinophidians are specialized burrowers, this evolutionary branch represents snakes' emergence from underground and readaptation to an above-ground existence. Apparently this development fostered rapid divergence into many new habitats and lifestyles

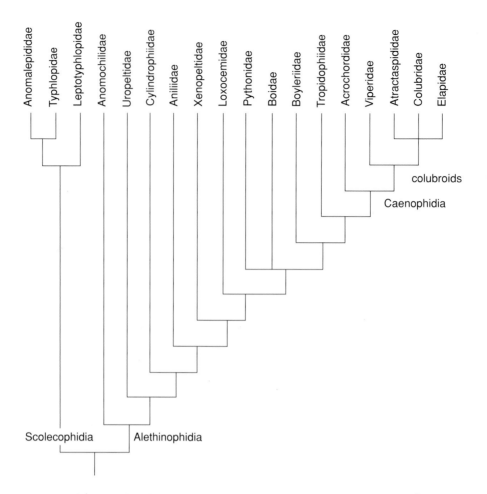

Figure 1.4. The assumed evolutionary relationships of living families of snakes. (After Zug, Vitt, and Caldwell 2001)

with the sequential divergence of approximately a dozen lineages. The final and most recent divergence, the caenophidian snakes gave rise to the great diversity seen in snakes today. The multiple lineages arising before the caenophidians have been called the henophidians, that is snakes with teeth on their premaxillae and remnants of a pelvic girdle, in contrast to the caenophidian snakes that lack teeth on the premaxillae and any evidence of a pelvic girdle. These two groups are often referred to as primitive and advanced (or modern), respectively, but this is a misrepresentation; the snakes in both groups are highly specialized in anatomy, physiology, and behavior.

The henophidians encompass an assortment of snakes commonly called boas and pythons. Not long ago, many of these families were assigned to the family Boidae. Our knowledge of evolutionary relationships now suggests that some boas, such as the Mascarene boas (Bolyeriidae) and neotropical dwarfboas (Tropidophiidae), are only distantly related to the true boas (Boidae) and pythons (Pythonidae). The pipesnakes (Aniliidae, Anomochilidae, Xenopeltidae) and shieldtails (Uropeltidae) are even more distantly related to the boas and pythons, and apparently to one another as well.

The caenophidians also include a number of lineages that diverged early. The wartsnakes (Acrochordidae)—not unexpectedly considering their strange appearance (loose skin and eyes on top of the head) and habits—are an early divergence, not closely related to any of the other families. The colubroids contain an assortment of families, which are related to one another, although the manner of the relationship is unclear. The most diverse family, Colubridae, is sometimes called a "trash-can group" because it contains many unrelated subgroups that are thrown together because biologists cannot decide where else to classify them. The complexities of potential relationships defy easy analysis. For example, the xenodontine snakes of the Western Hemisphere appear to share a common ancestor, but this lineage seems to have subdivided early into two groups that have followed similar patterns of diversification. Thus, certain species from different groups look very similar yet are only distantly related. Similar patterns of divergence and convergence have occurred in the colubrine, lamprophiine, and natricine snakes.

Of the three lineages of venomous snakes, biologists are reasonably confident that the vipers (Viperidae) represent an early divergence before the advent of the colubrids. Cobras, coralsnakes, and their allies (Elapidae) are offshoots of some colubrid snake, although we do not know which subgroup. The stilettovipers or molevipers (Atractaspididae) have at various times been assigned origins among the vipers, elapids, and certain African colubrid subgroups; for the moment, the consensus seems to favor the elapids. A fascinating feature of evolution within the elapids is the possibility that the seakraits (*Laticauda*) and seasnakes (*Hydrophis* and relatives) represent separate recent (perhaps within the last 6 million years) offshoots from the Australian elapid snakes.

The complexities and uncertainties of snake evolution make it a fascinating and challenging area of study. Many biologists are currently tackling this challenge with a variety of new tools, such as analysis of protein structure and mitochondrial DNA.

.2.

FOLKTALES

WHY SO MANY FALSEHOODS?

No other animal is the subject of as many myths and half-truths as the snake. In some cultures, snakes are associated with evil and sin; in others, snakes are symbols of healing or fertility. Of course snakes do not have magical or spiritual powers, and they are neither good nor bad in themselves. Human beings attribute symbolic and emotional meanings to snakes.

Much American and European folklore stems from the biblical allegory of the temptation of Eve by the serpent (Genesis 3:1–13). The cunning serpent tricks Eve into eating fruit from the forbidden tree, and persuading her mate Adam to do so as well. When God learns of the serpent's role, he places a curse on the serpent: "You will be punished for this; you alone of all the animals will crawl on your belly, and you will have to eat dust as long as you live. I will make you and woman hate each other; her offspring and yours will always be enemies. Her offspring will crush your head, and you will bite their heel." The Judeo-Christian tradition has long identified actual snakes with the biblical serpent, and snakes continue to arouse hostility. The redemptive power of religion has been depicted in legends in which the snake represents sin. One of the best known of these legends is that of Saint Patrick driving snakes out of Ireland. There are no snakes in Ireland because none ever reached its shores. Roughly 15,000 years ago, Ireland lay beneath a Pleistocene ice sheet. As the ice melted, the sea level rose and isolated Ireland from Britain before Ireland was warm enough to support snakes.

Many religions attribute mystical and supernatural powers to snakes. Folklore, which tends to address more concrete aspects of snake anatomy and behavior, portrays them both negatively and positively, depending on the culture.

Amazon treeboa (*Corallus hortulanus*)

69

WHAT SUPERNATURAL POWERS ARE ATTRIBUTED TO SNAKES?

In the biblical account of the serpent's temptation of Eve, the serpent can speak and reason. Elsewhere in the Bible, snakes are symbols of power. For instance, when Moses casts down his staff at the feet of the pharaoh, it turns into a snake that devours other snakes similarly created by Egyptian magicians. During the exodus from Egypt, God commands Moses to make a statue of a fiery serpent to protect any Israelites bitten by a fiery serpent (probably a reference to the parasitic guinea worm *Dracunculus medensis*, rather than a venomous snake). The mythical powers attributed to snakes in folklore probably arise in part from such religious sources. Even in Western culture, however, serpent symbolism is not always negative. The ancient Greeks revered snakes, and the modern symbol of the medical profession's gift of healing, the staff of Aesculapius, shows two snakes intertwined around the shaft.

Snakes' periodic shedding of their skin has also given rise to myths. Ancient Mesopotamians believed snakes to be immortal: each time a snake shed its skin, it was reborn. By contrast, some early Christians believed that snakes shed their skin in an attempt to remove the evil mantle placed on them by God.

Snakes have been worshipped in many cultures for protective powers and as bearers of fertility, probably because of their phallic associations. Native American tribes have left many signs of such beliefs, including sculptures of the "feathered serpent" symbolizing the Aztec god Quetzalcoatl.

Each year snake shamans of the Hopi Amerindians of Arizona perform an elaborate ritual snake dance to bring rain and to ensure crop fertility. The dance involves handling live snakes, including western rattlesnakes (*Crotalus viridis*), which represent revered ancestors. If honored, the ancestors will assure an adequate food crop and good fortune for the coming year. The captured snakes, sometimes numbering nearly a hundred, are placed in sacred clay jars and kept underground in a sacred vault until the time arrives to transport them to the village square.

After sundown the priests, in groups of three, perform a shuffling dance around the square. Eventually the leader of each group, the "carrier," is handed a snake. He places the snake in his mouth, holding it firmly between his teeth, and continues dancing. Eventually all the priests are holding snakes, whose heads are usually only inches from the priests' faces. Later the priests take up as many snakes as they can carry and run off in all directions to release the snakes outside the village. (Examination of the rattlesnakes after the dance has revealed that they have had their fangs removed, making them temporarily less dangerous.)

Fertility rituals in Africa and Southeast Asia also use snakes. In some West African tribes, young betrothed women join arms in a shuffling "python dance" in imitation of the African rock python (*Python sebae*). Participation in the dance is thought to

Cobras provided shade for the Buddha during his meditations of self discovery. Cobras are now revered by Buddhists and assume many sculptured guises in association with holy sites. (G. Zug)

bring fertility to their subsequent marriages. In Southeast Asia, one snake cult guarantees both human and crop fertility by having a young priestess kiss the head of an unrestrained wild king cobra (*Ophiophagus hannah*), a highly venomous snake.

WHICH TALES ARE TRUE?

Whenever outdoor people gather, snake tales are told. Like fish stories, they grow more grandiose with the telling, but somewhere underneath the ornamentation of the tale there may be an actual observation.

Are Snakes Cold and Slimy?

A snake can be cold to the touch if it is retrieved from a hibernaculum or caught beside a cold stream, but if caught in midsummer or when basking in the sun, it will feel as warm as a mammal. Snakes are ectotherms, and their body temperatures are

When the non-venomous green vinesnake (*Oxybelis fulgidus*) flattens its head, the larger triangular head enhances its threatening appearance.

dependent on their immediate surroundings (see *What Does It Mean to Be Cold-Blooded?*).

While a snake may be hot or cold to the touch, it will never be slimy. All reptiles have dry scaly skin, and snakes are no exception. Fish and amphibians, by contrast, have mucous glands in their skin; they feel slimy because mucus encases the entire body in a protective film. Reptiles' scales are a protective layer formed from keratin, the same material as human fingernails, mammalian hair, and bird feathers, none of which are slimy. The sliminess myth probably arose from a confusion between snakes and elongated and reduced-limbed amphibians and eels. Perhaps the iridescent color of some snakes is also mistakenly assumed to be wet and slimy.

Do Snakes Hypnotize Their Prey?

A snake slowly approaches a mother bird, which hops along a branch in agitation but does not fly away. The snake's steady gaze at the bird appears hypnotic, and the bird's apparent inability to fly away suggests that the snake's stare has a hypnotizing effect. But behind the female bird is her nest, and her apparent inability to escape is actually an attempt to distract the snake and protect her nest of eggs or fledglings. If she misjudges the snake's distance, she could become the snake's meal, but she is more likely to fly out of reach at the last moment and leave her nest undefended. Another possibility is that many animals recognize prey and predators by their movements, but fail to perceive a long thin snake slowly gliding toward them as dangerous.

Flattening and spreading the neck is a common defensive pose among colubrid snakes. This non-venomous aquatic smoothsnake (*Liophis typhlus*) creates a modest hood.

Do All Venomous Snakes Have Triangular Heads?

In the United States, a triangular head shape alerts humans that a snake is potentially venomous. All our pitvipers (rattlesnakes, copperheads, cottonmouths) have broad triangular heads and narrow necks, but so do some nonvenomous species such as watersnakes (*Nerodia*). And a nontriangular head does not necessarily signify lack of venom: the two United States coralsnakes (*Micruroides*, *Micrurus*), both venomous, have slender heads with little distinction between head and neck. Elsewhere, many venomous snakes (including most Elapidae and many rear-fanged colubrids) have small cone-shaped heads, and the nonvenomous pythons and boas have broad triangular heads.

Are All Hooded Snakes Venomous?

Spreading "hoods" (necks) are defense mechanisms. Only a few venomous species have hoods, notably the various cobras (*Naja*, *Ophiophagus*), and many nonvenomous snakes typically spread their hoods when disturbed or threatened. The North American hognose snakes (*Heterodon*) display broad hoods in defense when highly agitated, and strike with hood spread but usually with mouth closed.

Heterodon is mildly venomous, but its bite is less bothersome than a bee sting for most people (see *How Do Snakes Defend Themselves?*).

Do Injured Snakes Die before Sundown?

Time of day has no bearing on the death of any animal. A mortally wounded snake dies quickly, often instantaneously, just like a mammal or bird. However, nerve reflexes may cause muscle twitches for several hours after death, resulting in spasmodic movements of the trunk and jaws. This phenomenon is probably the origin of the superstition of death only at sundown.

Because of the lingering nerve reflexes, even a dead venomous snake is dangerous. An elderly Florida man died after a bite from the decapitated head of a canebrake rattlesnake (*Crotalus horridus atricaudatus*). While cleaning the snake to eat, he put his finger in the snake's mouth. That action stimulated a nervous reaction and the lethal bite.

Do Snakes Have Special Odors?

Contrary to popular belief, the North American copperheads (*Agkistrodon contortrix*) do not smell like cut cucumbers. We cannot account for the popular association of copperheads with the odor of cucumbers, a widespread belief in the eastern United States. All snakes have odors, as do all other animals, but in most the odor is barely discernible to the human nose. The one unpleasant exception is when they defensively spray their attacker with musk and feces.

Are Hoop Snakes For Real?

According to folklore, a hoop snake in the southern United States chases its victims by rolling after them. The hoop snake takes its tail in its mouth, forming a rigid circle. On a downhill chase, the hoop snake is said to build up speed and then fling itself like a spear, its poisonous spine-tipped tail felling the victim.

This vivid tale leaves many things unexplained. How does a snake lying on the ground turn itself on edge and begin to roll? How does the snake steer in the pursuit of its victim? The only accurate detail is the spine-tipped tail. A number of species, such as wormsnakes (*Carphophis*) and mudsnakes and rainbow snakes (*Farancia*) in the Southeast have pointed scales on the tips of their tails. These spines, though sharp, have no venom and are not dangerous.

Do Snakes Chase People?

The most outlandish stories of people being chased or stalked by venomous and nonvenomous snakes concern the American whipsnakes (*Masticophis*). The coachwhip supposedly uses normal serpentine locomotion to chase its human victim, then wraps the forepart of its body around the victim's legs. Once tripped, the victim is whipped to death with the coachwhip's long tail.

Coachwhips, like their close relatives the racers (*Coluber*), actively defend themselves and will at times advance toward a disturber if an escape route is not available. Other snake species, even the most "cowardly," can be provoked to advance on an adversary when they have no means of escape. Such behavior, manifesting the principle that "the best defense is a good offense," often drives away the snake's attacker.

Snake attacks also occur when a person stands between the snake and its hiding place. A British officer climbing a ravine in India looked up and saw a king cobra crawling rapidly toward him. Before he could escape, the cobra lunged at him—and dived into a hole between his feet. There are no authenticated reports of snake attacks or stalking that were not attempts at defense or escape.

Do Mother Snakes Swallow Their Young?

Many people believe that mother snakes swallow their young to protect them, and some even claim to have seen it. In thousands of hours of observation of snakes, in captivity and in the wild, no trained observer has seen this "protective" behavior. Nor have thousands of dissections revealed a single stomach full of baby snakes of the same species as the female dissected. Swallowing young would not be protective: snakes have efficient digestive systems, and anything that enters the esophagus soon arrives in the stomach and is bathed in acid and enzymes. Any small snake entering the digestive tract of a larger snake would be killed quickly through suffocation and digestion.

This tale probably arose and was reinforced by observation of a large snake eating a smaller snake, or by the discovery of near-term fetuses in a dead female's body.

Do Snakes Milk Cows?

In many pastoral societies throughout the world, it is a common superstition that snakes suck the milk from cows and goats. In North America, the milksnake (*Lampropeltis triangulum*) acquired its common name from the folklore that it could drain dry a cow.

Such beliefs are false. Cow and goat udders are sensitive, and snakes' mouths are characterized by tiny, sharply pointed teeth and inflexible lips. What cow is going to stand still when a snake clamps its mouth on her udder? And how can a snake suck milk when it cannot seal its mouth around the udder? There are other mechanical and behavioral implausibilities. A snake drinks by immersing its head, or at least its mouth, in water and then sucks in by expanding its body wall. Snakes also have only a small capacity for liquids and could not drain a cow dry. Finally, captive snakes spurn milk.

A long-nosed vinesnake (*Ahaetulla nasuta*) drinking from water trapped in the axilla of a plant. (K. Nemuras)

American milksnakes and other species are commonplace in pastures and around barns and cattle sheds where rodents are abundant. It is likely that snakes' close association with cows and goats, which occasionally go dry without explanation, gave rise to this superstition.

Do Venomous and Nonvenomous Snakes Interbreed?

Not all folklore is ancient in origin. In the late 1960s, a story began to circulate throughout the eastern United States that black-colored snakes (either *Coluber constrictor* or *Elaphe obsoleta*) mated with copperheads (*Agkistrodon contortrix*) or rattlesnakes (*Crotalus horridus*) to produce venomous young that looked like their nonvenomous parent. Newspaper reports and ill-informed naturalists gave this tale credence, reinforcing the prejudice that the only good snake is a dead one if you cannot trust appearances to tell you whether a snake is nonvenomous.

Although proponents of this tall tale cling to it, the laws of heredity prevent anything coming of such improbable crosses. For the same reason that the mating of a dog and a cat will not result in a litter of catogs or dogats, a mating between two different species of snakes will not produce offspring. Venomous rattlesnakes and harmless *Coluber* or *Elaphe* have diverged genetically for millions of years, yielding differences in chromosome number, structure, and content.

Only a distant or most cursory examination of a juvenile black racer (*Coluber constrictor* [left]) and an adult timber rattlesnake (*Crotalus horridus* [right]) could give rise to the rumor of interbreeding.

WHAT OTHER STRANGE TALES ARE TOLD ABOUT SNAKES?

In the Amish communities of eastern North America, people believe that snakes will not emerge in the spring until the first thunderstorm awakens them from winter dormancy. In reality, thunderstorms occur in all months of the year, making thunder a poor alarm clock. Slowly rising environmental temperatures "awaken" snakes and draw them out of their hibernacula in spring, and falling temperatures drive them underground in the fall.

It is said that some snakes whistle. Black racers and black ratsnakes can and do hiss, but the anatomy of their throats and mouths makes it improbable for them to whistle.

The fang-in-the-boot tale continues to be told around campfires. A man dies mysteriously. His son inherits his boots, and while wearing the boots for the first time also dies suddenly. When the boots are examined, a fang is found embedded in one heel. A good story, but the small amount of dried venom that could enter a scratch would at most produce inflammation, not death.

It is widely believed that, when walking in a "snaky" area, the last person in line is most likely to be bitten by a snake. In truth, a snake is as likely to bite the first person in line as the second or the last. It depends on whose foot intrudes into the snake's "personal space." The vibrations of an approaching walker warn most snakes away. In fact, more snakes see people than people see snakes, and it is more common to blunder into a wasp's or hornet's nest than to encounter a snake.

Snake charmers and their baskets with cobras are market-place attractions from North Africa to Southeast Asia. (M.D. Griffin)

CAN SNAKES BE CHARMED?

Snake charming is a traditional street entertainment in India, Bangladesh, Pakistan, and Myanmar, as well as in Egypt and other northern African countries.

In India, families of snake charmers hand down their skills and traditions from the men of one generation to the next. Indian snake charmers typically live in small rural villages. They catch cobras (*Naja*) in rat holes along the edges of village fields. Cobras with deformities, wounds, or poorly marked hoods, and those that are too aggressive, are usually rejected, as are those that would rather crawl away than defend themselves with raised hoods. The snake charmer then travels to a large city where tourists abound, such as Bombay, Calcutta, or New Delhi, or tours large villages or towns in the region. He finds a suitable spot on a street corner or in a market square.

The snakes are kept in wicker baskets or clay pots. An assistant often handles or displays a python or other large nonvenomous snake to attract a crowd while the charmer plays his flute. The snake charmer then removes the lid from the cobra's container. Contrary to popular belief, cobras are shy nonaggressive snakes that must be startled or provoked before they assume defensive postures. The sudden bright light usually startles and alarms the cobra, and it rises immediately. If the cobra does not react, the snake charmer sways the flute back and forth above the opening to entice the snake out.

The cobra extends above the basket opening in its full defensive posture, with hood fully spread and head facing the charmer. The snake usually does not extend more than 30 centimeters above the basket. Meanwhile the snake charmer slowly sways the flute in the cobra's face. The snake follows the motion of the flute, appearing to sway and dance to the music. If the flute were motionless, the cobra would retreat into its container or crawl away, but the moving flute threatens the snake. Sometimes the snake will strike at the flute. Some snake charmers lift the cobra and touch or kiss the top of its head as a climax to the show. When the charmer stops playing, the cobra sinks back into the container.

The illusion of cobra charming is that the snake hears the music and is entranced by its rhythm. But cobras, like other snakes, hear high-frequency airborne sounds poorly. In reality, they respond only to the motion of the flute, which they perceive as a danger. A cobra cannot strike upward, only downward—rather slowly, compared to a viper. Still, the snake charmer is in a dangerous position; he must be constantly alert to the snake's movements and behavioral eccentricities if he wishes to avoid a potentially fatal bite. When he touches or kisses the snake, he is in the snake's strike zone.

Are snake charmers' cobras really dangerous? Most snake charmers do perform with fully functional snakes. Some snake charmers, however, protect themselves by surgically removing all the fangs, tying shut the venom ducts, or sewing the lips closed. Others keep their animals intact but train them not to strike: the snake is repeatedly confronted with a flute-like object made of stone or metal; when the snake strikes, it experiences pain and eventually becomes conditioned not to strike. Hindu snake charmers take care not to harm their cobras—and they even release the snakes when they return to their home villages—because in their religion, cobras are sacred.

GIANT SNAKES: BIG AND BIGGEST

WHAT QUALIFIES AS A GIANT?

In theory, a snake is a giant if it, as a species, has an average size that is strikingly larger than the average of all other snake species combined. This is an impractical definition, however, for at least two reasons. First, what dimensions of size should be considered? We do not use weight in our definition, because weight data are unavailable for many species of snakes. And second, we would need to calculate the average length of all 2,800-and-some snake species, and for this we would need reliable data on each species. No such complete database exists, so the definition of "giant" must remain vague. In most snake species, individuals never exceed 1.25 meters, and many are much smaller. Species that reach lengths of over 2 meters are considered big snakes but not giants. Because everyone seems to agree that a snake longer than 3 meters is really big, that will be our lower limit for giant species of snakes.

ANACONDAS

The green anaconda or water boa (*Eunectes murinus*) is the best known of three species of anacondas, all of which live in South America. The green anaconda is found throughout northern South America, in the Amazon and Orinoco river basins and adjacent areas east of the Andes. Green anacondas are highly amphibious and spend much of their time in sluggish waters such as marshes, swamps, and slow-moving streams or backwaters of fast-flowing rivers.

Boa constrictor, *Boa constrictor*

Green anaconda (*Eunectes murinus*), maximum known length, 11.5 meters.

If the record length of 11.5 meters reported by Dunn is accurate (see Oliver 1958, Gilmore and Murphy 1993, and Murphy and Henderson 1997 for details), the longest individual snakes alive today are green anacondas. However, this maximum length is questioned by some biologists, who consider 9 to 9.5 meters to be the maximum length of the green anaconda. Most wild individuals never reach the maximum size. Six to 7 meters is the probable maximum length for most, and individuals over 5 meters are now rarely seen in the wild. If the green anaconda is not the longest living snakes, it certainly is the heaviest. A 5-meter-long individual weighs over 100 kilograms. One 5.8-meter-long individual measured nearly a meter in girth, roughly the girth of an 8-meter-long reticulated python. The other two species, the northeastern anaconda (*Eunectes deschaunseei*) from coastal French Guiana and northeastern Brazil and the yellow anaconda (*Eunectes notaeus*) from the upper reaches of the Paraguay River grow to 3 to 5 meters.

Anacondas are live-bearers, giving birth to baby snakes rather than laying eggs. The embryos develop in the females' oviducts or uteri, but there is no placenta. The developing embryos depend on the yolk stored in each egg, just as if the eggs were deposited externally. Litter size usually ranges from 20 to 70 newborns, and the newborns usually range in length from 68 to 80 centimeters, occasionally reaching nearly 1 meter. Growth in captivity indicates that juveniles grow rapidly and may reach 3 meters within 5 to 7 years.

Green anacondas are strongly but not exclusively aquatic. They bask on stream banks or on floating vegetation, and smaller individuals sometimes climb onto low branches or inclined tree trunks overhanging a stream. Hunting for prey is mainly

Reticulated python (*Python reticulatus*), maximum known length, 10.1 meters. (D.G. Barker, Vida Preciosa Intern., Inc.)

aquatic. Anacondas eat a variety of fish, amphibians, reptiles, mammals, and birds, ambushing terrestrial prey when it comes to water to drink or feed. The anaconda lies submerged; when the prey animal comes within striking distance, the snake strikes, bites, and grasps the prey and then quickly throws loops around the captured animal, constricting and killing it. Anacondas capture large animals such as caiman, peccaries, tapirs, and deer; thus they pose a possible (but unlikely) threat to humans. Reports of human–anaconda encounters are usually overblown.

RETICULATED PYTHON

The reticulated python (*Python reticulatus*) is the largest of the pythons, reaching a maximum verified length of 10.1 meters. Such a length is exceptional, however: reticulated pythons of 7.5 to 8.0 meters were frequently captured for zoos at midtwentieth century and earlier, but today few individuals as long as 5 or 6 meters are seen in the wild. Reticulated pythons are slender compared both to the other Asian giant, the Indian python, and to the green anaconda. They never reach the girth of these two species.

Reticulated pythons live in southeastern Asia from Myanmar (Burma) eastward to Vietnam and the Philippines and as far south as Indonesia. These giants probably once lived only in rain forests and transitional areas between forest and shrubby savanna, especially along riverbanks. They still inhabit forests, but have proved adaptable enough to invade agricultural areas and even suburbs and inner cities. In some river-port cities, they are common in wharf and warehouse districts, attracted by the abundant rats and many places to hide. In forests, reticulated pythons are often seen in trees; they are excellent climbers. They also take readily to water and are frequently seen swimming. In areas with sparse human populations, these pythons appear to be active both night and day but are more nocturnal in populated areas. When resting, they shelter in protected places such as mammal burrows, tree holes, hollow logs, and among the exposed roots of large trees.

Reticulated pythons are egg-layers. Females lay eggs from April to October. Young short females lay about 15 to 20 eggs in a clutch, but large females have been known to produce as many as 103 eggs in a single clutch. Females may coil around their eggs, but neither the female's nor the eggs' temperatures are raised. The eggs hatch in 55 to 80 days, producing hatchlings 60 to 75 centimeters long.

Reticulated pythons capture prey by slow stalking or ambush. Their prey includes a variety of mammals (small deer, rodents, pigs, cats, dogs, mongooses, civets), birds (including chickens and ducks), and lizards. Big reticulated pythons can swallow large prey. A python is said to have swallowed a 25-kilogram pig being prepared for butchering. A 14-year-old Malay boy was grabbed, constricted, and swallowed by a 5.2-meter python. Nevertheless, some snake hobbyists with small children allow pythons to roam freely about their homes.

AFRICAN ROCK PYTHONS

The African rock pythons (*Python natalensis* and *Python sebae*) are the largest snake of Africa. Two of the four truly gigantic snakes in the world, they are surpassed in length only by the green anaconda and the reticulated python and, like these two giants, the rock pythons are thick-bodied, heavy snakes. Of the two, the Central African rock python (*P. sebae*) has a record length—questioned by some—of 9.45 meters (Duncan 1847). Doubters note that even the 7.5-meter individual captured at a biological station is unconfirmed; that is, there is no photograph or specimen voucher in a museum. Confirmed lengths of 5.5 meters exist for both species. Today, few individuals of either species are found near the maximum length, although lengths of 2.8 to 4.0 meters are common, and larger individuals are seen occasionally.

Rock pythons live in grasslands and savannas with dense vegetation, often near streams, lakes, swamps, or marshes. They seem to require high humidity and are seldom found in arid habitats or deserts. Rock pythons were once common in most of Africa, from the southern limits of the Sahara southward to South Africa, but they have become increasingly scarce.

During the dry season, they typically estivate in hollow trees or logs, or underground in animal burrows. These large snakes are often seen prowling the vegetation on the banks of a waterway. Good swimmers, they take readily to water,

Rock python
(*Python natalensis*), maximum
known length,
6.5 meters.

especially in the hottest weather. They climb trees, but less commonly than some other pythons and boas.

Rock pythons tend to be most active in the morning and evening when temperatures are moderate, capturing prey by active search and ambush. The snake ambushes prey by hiding in thick vegetation beside a game trail or on the low branches of shrubs or trees overhanging the trail. At a watering hole or where prey cross a stream, the python typically rests on the bottom with only its head at the water's surface. With an explosive strike, it bites and holds the prey, then coils around and constricts it. Rock pythons eat a wide variety of mammals, including the young of large species (duiker, gazelle, steinbok), small antelopes, leopards, hares, rodents (including porcupines), jackals, monkeys, and various birds from doves to cranes and geese. Young rock pythons may eat frogs and small birds and mammals. Large pythons can eat immense meals: one swallowed three jackals, and a 4.8-meter rock python ate an adult impala. After large meals, pythons often seclude themselves and bask for several days; they may not eat again for weeks and are capable of fasting for months. One captive went 29 months between feedings.

Gravid females (oviparous females carrying eggs before they are laid) find concealed sites, such as termite mounds, hollow trees, stumps or logs, rock crevices, dense underbrush, or abandoned mammal burrows, for egg deposition. The larger the female, the more eggs she can lay. Most clutches contain 30 to 50 eggs, but may consist of as few as 16 eggs. Larger clutches are possible; a female at the London Zoo laid a clutch of about 100 eggs. Although females coil around their eggs for the duration of incubation, which lasts from about 50 to 100 days, they do not seem to be brooding. Young rock pythons are 50 to 70 centimeters long at hatching. They grow rapidly and can reach lengths of 1.2 to 1.4 meters in 1 year.

Unprovoked attacks on humans are rare, although several cases of predation on humans have been confirmed. Normally rock pythons remain calm when discovered; if given the chance, they crawl away. If agitated, they hiss and then strike with the power of a boxer's punch. They are nonvenomous, but their large teeth can inflict deep painful lacerations.

INDIAN PYTHON

The Indian python (*Python molurus*) is the most popular giant of the pet trade. These pythons have handsome patterning and are usually docile. Their record length is 5.8 meters, but individuals over 4.0 meters are uncommon. The Burmese subspecies supposedly produces the largest individuals, including some reportedly longer than 6 meters, but we have been unable to confirm these lengths.

Indian pythons are not confined to India; they occur from the Indus River of Pakistan eastward through India to southern China, and southward to Sumatra and Java. Because of human persecution and habitat modification, however, they are now rare or extinct in many areas. The pet and leather trades have depleted many wild populations of this snake. Although the Indian python is legally protected in some countries, such as India, poaching continues to endanger populations. Indian pythons occur in a variety of habitats from coastal mangrove forest to montane forest at elevations above 2,400 meters. They prefer forest habitats, particularly scrub forest with thick canopy, but persist in marshlands near urban sprawl and occasionally even in urban storm-sewer systems.

These pythons rest in tree and rock cavities, often sharing the burrows of diggers such as the Indian porcupine. Occasionally several individuals den together in such resting sites; they often bask near the entrances of their lairs. Smaller individuals climb trees, and Indian pythons of all sizes readily swim.

Between February and June, female Indian pythons seek secluded lairs to lay their clutches of 12 to 54 eggs. The female coils around the eggs and broods them like a bird, elevating her body temperature to provide the elevated incubation temperature necessary for normal development of Indian python embryos. The female raises her body temperature by means of minute muscular contractions similar to shivering in a mammal, although without the violent shaking (see *What Does It Mean to Be Cold-Blooded?*). Like most reptiles, pythons lack effective body insulation. Coiling reduces the female's exposed surface area and rate of heat loss; sheltered lairs offer temperature stability and some insulation. Thus, a female can maintain a clutch temperature 6 to 8°C above that of her surroundings. Incubation

Indian python (*Python molurus*), maximum known length, 5.8 meters.

lasts 55 to 80 days even with brooding. Hatchlings are 43 to 74 centimeters in total length.

Indian pythons are active both day and night, searching for food or mates or moving to new feeding or resting sites. Their main prey are mammals, including small deer, jackals, dogs, monkeys, porcupines, squirrels, rats, hares, mongooses, and domestic goats. Birds, including chickens, are also eaten, as are frogs and fish. One 5.5-meter python captured and swallowed a 1.27-meter leopard. Stories abound of pythons devouring babies and small children. Although a large Indian python could do so, there are no authentic records of such occurrences.

AUSTRALIAN SCRUB PYTHON

The amethystine or scrub python (*Morelia amethystina*) is a native of Australia, New Guinea, and Indonesia. A dozen or more species of pythons live in this region, but the scrub python, a recently discovered relative, and the carpet python (*Morelia spilota*), are the only giants among them. One scrub python measured 8.5 meters in total length, but most large individuals are 3 to 4 meters long. The average adult female is about 2.4 meters, the average adult male about 0.5 meter smaller. Residents of Arnhem Land in northeast Australia claim that the Oenpelli python (*Morelia oenpelliensis*) is even longer than the scrub python, yet biologists can

Scrub python (*Morelia amethystina*), maximum known length, 8.5 meters.

substantiate lengths of only about 2 meters for this species. Carpet pythons occur broadly throughout Australia and southern New Guinea. They display a variety of color patterns and in most populations, maximum length is less than 3 meters.

Scrub pythons live in a variety of habitats, from rain forests to open savanna. They are frequently seen along the banks of streams and lakes; in areas with cool winter temperatures, they aggregate in rocky ravines for hibernation. They regularly bask, and several individuals often catch the morning rays together on a rock outcrop or stream bank.

Females lay 7 to 20 eggs. Some, perhaps all, coil around their egg clutches throughout the incubation period. Nest sites are often hollow logs or stumps. Hatchling scrub pythons are 45 to 60 centimeters in total length.

Scrub pythons hunt chiefly at night. Like other pythons, they subdue their prey by constriction. They appear to feed exclusively or nearly so on mammals, and are capable of catching and swallowing small kangaroos and wallabies. This is particularly remarkable because, unlike the other giant pythons, scrub pythons are slender snakes. A 3- to 4-meter scrub python has a girth of less than 40 centimeters.

BOA CONSTRICTOR

The boa constrictor—one of the few animals and plants whose scientific name *Boa constrictor* is the same as its common name—is the best known of the giant constrictors, at least by name. Many people equate all giant snakes with boa constrictors. True boas, including anacondas, are live-bearers and anatomically distinct from pythons.

These boas are the smallest of the boa–python giants, with a maximum reported size of 4.2 meters. The 5.6-meter boa constrictor supposedly shot by C. F. Pitten-

Boa constrictor (*Boa constrictor*), maximum known length, 4.2 meters. (C. H. Ernst)

drigh in Trinidad is a case of mistaken identity; the snake was actually an anaconda. The largest boa constrictors come from northern Venezuela and Trinidad; elsewhere individuals over 3.0 to 3.5 meters are rare.

The boa constrictor ranges from coastal northern Mexico southward through Central America to northern Argentina. It lives in a wide variety of habitats, including tropical rain forests and jungles (both evergreen and deciduous), brushy and grassy savannas, rocky semideserts, and cultivated plantations and fields. Boas have not proved as adaptable as Indian and reticulated pythons to urban life.

Almost any sheltered crevice or burrow will serve as a daytime retreat. Boas do most of their foraging for food and searching for mates at night. Young small boas are good climbers and search for prey in trees. Older and larger boas usually forage only at ground level, often lying among leaves on the forest floor, camouflaged by their pattern, waiting for prey to wander by. They eat almost any small mammal (opossums, all types of rats and mice, agoutis, and even small carnivores such as ocelots), birds, and lizards. Unlike the anacondas and large pythons, boa constrictors seldom enter water.

The timing of reproductive activity varies with locality, but matches birth with the most equitable season for the young to survive. Litters of 21 to 64 young are born after gestation periods of 119 to 295 days. Newborns average about 50 centimeters in total length.

ASIAN RATSNAKES

The largest snakes in their respective groups are all native to Asia: the reticulated python, the king cobra, and the Asian ratsnakes (*Ptyas*, the largest nonvenomous snakes in the family Colubridae). Two species of Asian ratsnakes attain giant status. The common Indian banded ratsnake (*Ptyas mucosus*) has a record length of 3.6 meters, and the keeled ratsnake (*Ptyas carinatus*) is slightly longer with a record length of 3.7 meters. More typically, adults of these two species average 2.0 to 2.5 meters, males growing somewhat larger than females. A third, slightly smaller species, the Indochinese ratsnake (*Ptyas korros*), has adults of 1.8 to 2.5 meters.

Indian banded ratsnakes range from Turkestan and Afghanistan eastward through India and Sri Lanka to southwestern China and southward through Vietnam and Thailand to Malaysia. This species prefers open country with high grass and is often seen in rice fields and around grain-storage buildings where it hunts for rodents. Indian ratsnakes are good climbers and swimmers.

A diurnal snake, it actively searches for prey. Young Indian ratsnakes eat mainly frogs; within a year, they switch to mice and rats. Elsewhere, ratsnakes of all ages live mainly on frogs and toads. Lizards, snakes, small turtles, and bats are also occasional prey. Prey is swallowed alive, squeezed to death in the jaws, or pressed to death against a hard object—a behavior also characteristic of the North American bullsnakes (*Pituophis melanoleucus*) and the foxsnakes (*Elaphe vulpina*). Asian ratsnakes do not constrict prey within body coils as do the *Elaphe* ratsnakes.

Six to 16 eggs are laid between March and September. Eggs hatch in about 60 days and yield young 32 to 47 centimeters long.

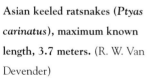

Asian keeled ratsnakes (*Ptyas carinatus*), maximum known length, 3.7 meters. (R. W. Van Devender)

The Indian ratsnake is an excitable species, even after long periods in captivity. When aroused, it coils, raises and arches its forebody, inflates its throat, hisses loudly, and strikes repeatedly.

The keeled ratsnake—sometimes assigned to the genus *Zaocys*—is found from Thailand southeastward through Indonesia to the Philippines. It is a forest dweller and oviparous, but little else is known of its life history. The Indochinese ratsnake shares a similarly broad range and habitat preference with the Indian banded ratsnake.

Asian ratsnakes have suffered heavily from exploitation for the leather trade.

KING COBRA

The king cobra or hamadryad (*Ophiophagus hannah*) is the world's largest venomous snake. These giant cobras are found throughout tropical Asia, from India eastward to Vietnam, southern China, and the Philippines, and southward through Malaysia and Indonesia. Hamadryads are not common anywhere, and are now considered rare in India. They are usually found in forests, often near streams, in mangrove swamps, and on tea and coffee plantations. They prefer habitats with heavy rainfall and dense undergrowth. King cobras readily ascend shrubs and low trees. Adults attain lengths of about 3 meters; only a few individuals grow larger. The largest hamadryads reside in the Malay Peninsula, and the maximum verified length is 5.58 meters (Aagaard 1924).

The king cobra's generic name (*ophis*, snake; *phagein*, to eat) indicates its dietary preference: other snakes. Although it preys predominantly on nonvenomous species (including small pythons), king cobras also eat highly venomous snakes, such as cobras (*Naja*) and kraits (*Bungarus*). King cobras are cannibalistic and eat smaller individuals of their own species. They also prey on lizards, including large monitor lizards (*Varanus*). Active foragers, they hunt during the day.

King cobras mate in the late dry season, between January and March; the female lays a clutch of 20 to 50 eggs in April, May, or June. The king cobra is the only snake known to construct a nest in which to incubate its eggs. The nest consists of an untidy heap of leaves, twigs, and other vegetation, which the female scrapes together with her coils. The eggs are deposited and covered, presumably benefiting from the heat generated by the decaying vegetation. Above them, the female hollows out a second chamber where she or her mate rests, guarding the eggs against predators. The other adult often remains nearby, but it is not known whether the male and female alternate guard duty. After 60 to 90 days, hatchlings 50 to 53 centimeters long emerge from the eggs. There is no evidence of any further parental care.

Adult king cobras (*Ophiophagus hannah*), maximum known length, 5.6 meters.

A favorite nest site for king cobras is the dead leaves that accumulate beneath clumps of bamboo. The nest looks no different than other heaps of dead bamboo leaves and is easily overlooked; the resident adult remains quietly coiled unless provoked. H. C. Smith recounts that "14 people accompanied by 7 dogs twice passed at different times within two yards of the nest, and yet the Hamadryad failed to show itself, and the nest remained undiscovered until I prodded the heap of leaves with a small cane."

Despite its massive size and fierce reputation, the king cobra is not an aggressive snake. Much more deliberate in its actions than the smaller, excitable cobras (*Naja*), it will quickly crawl away if given the opportunity. If cornered or attacked, however, it readily defends itself. Under attack, the king cobra coils, raises its forebody to a height approximating one-third of its body length—the world's largest individual would have extended about 1.8 meters!—flattens its neck into a long narrow hood, and methodically strikes downward at its foe. Its venom is not as toxic as that of the common Indian cobra or krait, but because of its large venom glands it can inject a greater quantity of venom. Whitaker reported that these glands can contain as much as 6 milliliters of venom, enough to kill an Asian elephant. Fortunately, bites to humans are rare.

TAIPANS

Australia has many more venomous snake species than nonvenomous ones. All of these venomous snakes are elapids, distantly related to the cobras, kraits, and American coralsnakes. Two species, the common taipan (*Oxyuranus scutellatus*)

Common taipan (*Oxyuranus scutellatus*), maximum known length, 3.4 meters. (C. A. Ross)

and the inland taipan (*Oxyuranus microlepidotus*), attain giant status. Both are tropical snakes, living in northern and northeastern Australia and southern New Guinea. The inland taipan, as its name indicates, inhabits the drier Australian interior, largely restricted to the Simpson Desert. The common taipan occurs broadly elsewhere in Australia and New Guinea.

The common taipan is the longest Australian elapid, at a maximum total length of 3.4 meters. This record length makes it the second-longest venomous snake in the world, although considerably smaller than the king cobra. Because taipan venom is far more toxic than king-cobra venom, this giant is as threatening as the king cobras. The record length for the inland taipan, 2.5 meters, is probably exceeded by many individuals, given that adult taipans of both species commonly reach about 2 meters. Females are larger than males among the inland taipans; males are larger among the common taipan.

Taipans are egg-layers. Females produce clutches of 3 to 25 eggs, usually in November and December. Incubation lasts 10 to 14 weeks, and hatchlings are 30 to 35 centimeters long.

Taipans are active hunters, mainly diurnal. Mammals appear to be the exclusive prey of the inland taipan and the principal prey of the common taipan. In a confrontation with humans, these snakes aggressively defend themselves with rapid and repeated strikes. A large individual on the defense is awesome, especially if the observer is aware of the high toxicity of their venom.

BUSHMASTERS

The bushmasters (*Lachesis*) are the largest venomous snake in the Americas and the largest vipers in the world. Three, perhaps four, species occur in tropical America: Central American bushmaster (*Lachesis stenophrys*), South American

Central American bushmaster (*Lachesis stenophrys*), maximum known length, 3.8 meters.

bushmaster (*L. muta*), black-headed bushmaster (*L. melanocephala*), and the Darien bushmaster (specific status uncertain). In all species, 2- to 2.5-meter bushmasters are not uncommon, but it is rare to see any of these well-camouflaged snakes on leaf-littered forest floors. The record length of 3.75 meters is from an individual of the Central American bushmasters, and there are unconfirmed reports of the South American bushmaster reaching 4 meters.

Primarily forest dwellers, all species of *Lachesis* inhabit moist to wet tropical forests to elevations of at least 1,500 meters. Bushmasters range from southern Nicaragua southward into northern South America; they are also found on Trinidad.

Bushmasters are active mainly at twilight and at night. As ambush hunters, they remain stationary for days at a time. They can spend as many as 25 days at one site, and the total home range of this large snake may be quite small. A wild female Central American bushmaster radio-tracked for 35 days was alert only 40% of the time and moved actively only 1% of her alert time.

The bushmaster's main prey is small mammals, especially spiny rats and opossums, but they also capture small birds and frogs. Although bushmasters occasionally capture prey while moving, most prey is captured from ambush. After selecting a sheltered site along a rodent or opossum trail, they wait. Ambush sites are selected by prey odor; prey detection and capture depend on sight (visual and infrared).

San Francisco gartersnake
(*Thamnophis sirtalis
tetrataenia*), west coast
North America

Timor scrub python (*Liasis timorensis*). Timor, Lesser Sunda Islands

Chinese red-headed ratsnake (*Elaphe moellendorfei*), southeastern China

Juvenile king cobra (*Ophiophagus hannah*), South Asia

Irregular smoothsnake
(*Liophis anomala*), South
America

Rainbow snake (*Farancia
abacura*), southeastern
North America

Malagasy treesnake
(*Stenophis* cf. *beatiana*),
Madagascar

Dwarfboa (*Ungaliophis continentalis*), northern Middle America

Wagler's treeviper (*Tropidolaemus wagleri*), Indomalaya

Black-necked gartersnake (*Thamnophis cyrtopsis*), western North America

Texas ratsnake (*Elaphe obsoleta lindheimeri*), south-central North America; this individual lacks scales on most of its body

Tentacled snake (*Erpeton tentaculatum*), Indomalaya

Common deathadder (*Acanthophis antarcticus*), tropical Australia

Water python
(*Liasis fuscus*)
Northern Australia and
New Guinea (D. Barker)

Western brownsnake
(*Pseudonaja nuchalis*),
Australia wide, except
eastern seaboard

Honduras milksnake
(*Lampropeltis triangulum
hondurensis*), Honduras,
Central America

Spiny bushviper (*Atheris hispida*), central Africa

Rhinoceros viper (*Bitis nasicornis*), sub-Saharan Africa

Everglades ratsnake (*Elaphe obsoleta rossalleni*), southern Florida, North America

Cape birdsnake (*Thelotornis capensis*), South Africa

Western Children's python (*Liasis perthensis*), western Australia

When the prey passes by, the unseen snake strikes and envenomates it. The prey moves on but soon dies. The snake leisurely follows the odor trail to the prey, determines that it is dead, and swallows it. After feeding, a bushmaster seeks a secluded site, such as a hollow log, dense vegetation, or the exposed roots of a tree, and rests until its meal is digested. It may emerge briefly to bask, possibly to enhance digestion. If its prey is sufficiently large, a bushmaster can thrive on only 6 or 7 meals a year.

The bushmasters are the only New World pitviper to lay eggs. Clutches contain 9 to 16 eggs. When a female finishes laying the eggs, she coils around them and maintains her protective vigil, although not brooding, for 60 to nearly 90 days. Hatchlings are about 30 centimeters long.

MAMBAS

African mambas (*Dendroaspis*) are feared for their reputedly fierce disposition, deadly venom, and size. There are 4 species of *Dendroaspis*, a black mamba and 3 green mambas. The black mamba (*Dendroaspis polylepis*) is a terrestrial snake. It commonly reaches 2.5 to 3.0 meters and has a maximum recorded total length of 4.5 meters. The common mamba (*D. angusticeps*) is an arboreal snake, smaller than the black mamba, with a maximum length of 2.5 meters and an average size of 1.8 meters.

The two snakes live in different but overlapping habitats. The black mamba inhabits much of the southern half of tropical Africa, from southern Kenya and Zaire southward; the common mamba is found in eastern Africa from Kenya south

Black mamba (*Dendroaspis polylepis*), maximum known length, 4.5 meters.

through Natal. The black mamba lives in open bush country, generally below 1,500 meters elevation, especially around granite hillocks (kopjes), and in riverine forests. It occupies a permanent den, usually in a termite mound, rock crevice, or hollow tree, to which it returns each night. Often it basks near the entrance to its retreat. Black mambas are equally adept on the ground and in trees, but they are mainly terrestrial snakes.

The common mamba is almost completely arboreal, seldom descending to the ground. At night, common mambas retreat to tree hollows or clumps of leaves. They are typically residents of thick evergreen coastal forests, but also inhabit bamboo thickets, mangrove forests, and tea plantations.

The temperaments of the two snakes differ strikingly. The black mamba is very excitable; when provoked or cornered, it strikes quickly and often. In contrast, the common mamba is shy and unaggressive. Both snakes will flee if given the chance, but if a black mamba perceives itself as trapped, it coils, gapes, protrudes its tongue (displaying the black mouth lining from which it derives its common name), spreads its neck into a narrow hood, hisses, and strikes repeatedly. These strikes may be so violent as to lift 40% (typically 25 to 30%) of the snake's body off the ground. This seemingly aggressive behavior has made the black mamba the most feared snake in Africa. The common mamba also actively defends itself, but its defense is less vigorous, and it does not spread its neck, gape, or protrude its tongue.

Both species are active diurnal hunters. The black mamba forages mainly at ground level, with as much as one-third of its forebody raised off the ground. The common mamba crawls across a discontinuous platform of branches and twigs in search of prey, but may also lie quietly among the leaves and ambush its prey. The black mamba eats small mammals (rats, mice, shrews, bats, rock hyraxes), birds, and lizards. The common mamba takes birds (all stages from eggs to adults), small arboreal mammals, and lizards. Both species strike their prey and usually immediately release it, allowing their fast-acting venom to do its work.

Like other terrestrial elapids, mambas are oviparous. The black mamba lays 6 to 18 eggs, the common mamba 6 to 17 eggs. Hatchling black mambas are 40 to 60 centimeters long and those of its arboreal cousin 35 to 45 centimeters. The common mamba usually mates among tree branches, the black mamba on the ground or in its den.

OTHER GIANTS AND NEAR-GIANTS

Although no North American snake meets our criteria for giants, many nonexperts consider some of them giants. Let us briefly survey these North American species and a few other large snakes.

Indigo snake (*Dry-marchon corais*), maximum known length, 2.6 meters.

The indigo snake (*Drymarchon corais*) is the largest snake found north of Mexico, with a record length of 2.63 meters. This gentle and nonvenomous near-giant inhabits southern Alabama, Georgia, and Florida, but its main distribution is from southern Texas and Sonora to central South America. The indigo snake is diurnal, foraging mostly before noon. It is not a constrictor; instead it grabs its prey and begins to swallow it immediately. Prey consists of a variety of snakes, lizards, small turtles, frogs, toads, small mammals, and ground-nesting birds; some tropical populations even prey on fishes. Snakes, including venomous species, seem to be a preferred prey, and the indigo snake is somewhat immune to their bites. Venomous snakes are usually seized by the head and chewed vigorously until immobilized.

A few other North American snakes are near-giants. The common coachwhip (*Masticophis flagellum*) of the southern United States has a record total length of 2.59 meters, and the black ratsnake (*Elaphe obsoleta obsoleta*) of the eastern United States reaches a maximum of 2.57 meters. The black racer (*Coluber constrictor constrictor*) is often confused with the black ratsnake, but does not grow as large; its maximum length is only 1.9 meters. These three nonvenomous snakes are daytime prowlers and active hunters. Birds and mammals, particularly rodents, are prime prey for the coachwhip and black ratsnake, although they eat other prey from insects to lizards. The black racer has a more catholic diet of invertebrate and vertebrate prey. All three are slender and crawl rapidly when pursued. They cannot outrun a human, but they easily elude capture.

The two largest venomous snakes in North America are the eastern (*Crotalus adamanteus*) and western (*C. atrox*) diamondback rattlesnakes. They share a diamond-like pattern but are not closely related. The eastern diamondback is the largest

Eastern diamondback rattlesnake (*Crotalus adamanteus*), maximum known length, 2.5 meters.

venomous snake north of southern Mexico, with a record of 2.51 meters. Few adults, however, are longer than 1 to 1.5 meters. The eastern diamondback is also one of the heaviest snakes in North America, by far the most bulky of the venomous snakes in the United States. The western diamondback has a record of 2.34 meters. Adults over 1.2 meters are common but few ever attain 1.5 meters.

These two rattlesnakes reside in the opposite southern corners of North America, the eastern diamondback in the Southeast and the western one in the Southwest. In the Southeast, the eastern diamondback is most common in dry lowland palmetto, pine-wiregrass flatwoods, and pine–turkey oak woodlands. The western diamondback lives in a greater variety of habitats, ranging from the scrub forest and grasslands of eastern Texas to the more arid habitats of Arizona and northern Mexico.

Surprisingly, the biology of neither species is known particularly well. Rattlesnakes tend to be nocturnal hunters, like their mammalian prey. Adult eastern diamondbacks are often seen in the early morning or evening when rabbits, their favorite prey, are active, but they hunt at night as well. Juveniles feed chiefly on mice and small rats, and adults also eat cotton rats, white-footed mice, squirrels, and some birds. Western diamondbacks eat mice, rats, squirrels (both tree and ground), kangaroo rats, pocket gophers, cottontail rabbits, and jackrabbits, but they also prey on birds, snakes, and lizards. Both diamondbacks use ambush and actively follow scent trails.

For a large venomous snake, the eastern diamondback has an almost docile temperament. Because it usually lies quietly coiled when first discovered, it has been called "a gentleman among snakes." However, eastern diamondbacks strike quickly if provoked. Although their venom is only mildly toxic, the great quantity delivered in each bite makes them dangerous.

Although surpassed in both length and bulk by its eastern cousin, the western diamondback is nervous and easily aroused. A large, irate western diamondback is an awesome foe: it coils, raises its head and neck as much as 50 centimeters above its coils, continually rattles, strikes, and occasionally moves toward its adversary. Its venom is more potent than that of the eastern diamondback, and it injects large quantities when it bites.

Among the Central and South American pitvipers, only the bushmasters and the terciopelo (*Bothrops asper*) are as large as the diamondback rattlesnakes. The terciopelo has a verified maximum total length of 2.5 meters. Large individuals are heavy and thick-bodied. Most adults range from 1.2 to 1.8 meters. Terciopelos have a broad distribution, from southern Mexico to northern South America as far east as Trinidad. Only the fer-de-lance (*Bothrops atrox*), the jararacussu (*Bothrops jararacussu*) of southern Brazil, the Saint Lucia lancehead (*Bothrops caribbaeus*), and the Martinique lancehead (*Bothrops lanceolatus*) approach the terciopelos in size, with maximum reported lengths greater than 2 meters. The cascabel or neotropical rattlesnake (*Crotalus durissus*) has a maximum length of about 1.8 meters. Only two coralsnakes (*Micrurus spixii*, M. *surinamensis*) can be considered large, and their maximum confirmed lengths are less than 1.8 meters.

In Africa, five cobras qualify as near-giant venomous snakes. The African water cobras (*Boulengerina annulata*, B. *christyi*) have a maximum total length of about 2.5 to 2.7 meters. These cobras are semiaquatic, living among the shore and water vegetation of the rivers and lakes of central Africa. They are predominantly amphibian and fish predators.

The black-necked cobra (*Naja nigricollis*) and the forest cobra (*Naja melanoleuca*) qualify as near-giants with record lengths of 2.8 meters and 2.7 meters, respectively. The Egyptian cobra (*Naja haje*) is smaller, with a maximum size of about 2.5 meters. All three are terrestrial, although they occasionally forage along waterways for cold-blooded prey, such as frogs, lizards, and other snakes.

Africa has no other lengthy snakes. The arboreal boomslang (*Dispholidus typus*) and Kirtland's birdsnake (*Thelotornis kirtlandii*) have maximum lengths of 1.6 to 1.8

African black-necked spitting cobra (*Naja nigricollis*), maximum known length, 2.7 meters.

meters. A few other colubrid snakes occasionally reach 1.5 meters, but most have maximum lengths less than 1 meter. The Gaboon viper (*Bitis gabonica*) has a record length of 1.8 meters; owing to the species' heavy-bodied physique, this particular individual probably had the girth of a 4- to 5-meter Indian python. Although somewhat shorter at 1.2 meters, the rhinoceros viper (*Bitis nasicornis*) is similarly heavy-bodied.

Temperate Eurasia lacks giant snakes and has only a few near-giants. Several ratsnakes (*Elaphe*) reach total lengths of more than 2 meters. The largest ratsnake is the four-lined ratsnake (*E. quatuorlineata*) of central Europe, with a record length of 2.5 meters. The long whipsnake (*Coluber jugularis*) of Southwest Asia reaches lengths near 3 meters. None of the Eurasian vipers (*Vipera*) is large, and most are less than 1 meter as adults.

Tropical Asia has many giants but few near-giants. Most Asian snakes are less than 1.5 meters long. An occasional ratsnake reaches 2 meters, such as the copperhead ratsnake (*Elaphe radiata*) and the red-tailed racer (*Gonyosoma oxycephala*). Some bronzebacks (*Dendrelaphis*) and whipsnakes (*Ahaetulla*) appear longer than they are, owing to their slender bodies and pointed heads, but few individuals exceed 1.5 meters in total length and most are less than 1 meter. The same is true of most of the venomous cobras (*Naja*), kraits (*Bungarus*), vipers (*Vipera*), and pitvipers (*Agkistrodon, Calloselasma, Deinagkistrodon, Trimeresurus*).

Gaboon viper (*Bitis gabonica*), maximum known length, 1.8 meters.

The Asian genus *Boiga* (catsnakes or cat-eyed treesnakes) contains a few species of large snakes. The dogtooth catsnake (*Boiga cynodon*), for example, has a maximum length of 2.8 meters. The most notorious of these snakes is the brown treesnake (*Boiga irregularis*). Its natural home is New Guinea and the Solomon Islands, but the brown treesnake is now pervasive on the Pacific island of Guam. Apparently it was accidentally carried to Guam at the end of World War II when the U.S. armed forces moved supplies from New Guinea. Its presence in Guam, unrecognized until the early 1950s, was largely ignored. By the early 1980s, it could no longer be overlooked: the brown treesnake had literally eaten all the birds on the island. A single species of snake had eliminated two dozen bird species from a single island!

No location is safe from this long, skinny, rear-fanged snake. It forages on the ground and is an excellent climber. It is found everywhere from the centers of towns to the middle of the forest, and because it has no natural enemies, its numbers are immense (16 to 50 individuals per hectare). The brown treesnake has even decimated the lizard population of Guam; lizards are the preferred food of the young snakes. Because it occupies warehouses and wharves and is frequently transported with cargo, it is a serious threat to all other Pacific islands.

Brown treesnakes reach a maximum length of 2.3 meters in native areas and 3.1 meters on Guam. They appear longer because of their extremely slender bodies and proportionately large heads. Typically unexcitable, they move away from any

Brown treesnake (*Boiga irregularis*), maximum known length, 2.8 meters. (G. H. Rodda)

disturbance. But they do defend themselves and, being rear-fanged, they are potentially dangerous although not lethal. There are several reports of this snake biting sleeping babies; presumably the snakes mistook the fluttering eyelids of sleeping children for prey. No children have died, but hospital care was necessary. For an adult, the brown treesnake's venom is mild and may cause only minor local swelling, itching, and bleeding.

Further south in New Guinea and Australia, the olive python (*Liasis olivaceus*) is reported to reach a maximum total length of about 7 meters; however, such lengths are unsubstantiated and olive pythons more than 3 meters long are rare. Most adult olive pythons range between 1.5 and 2.0 meters long. Like their scrub python cousins, olive pythons are slender; a 2-meter-long individual has a girth no larger than the wrist of an adult woman. The record length for the tigersnake is about 2 meters, but adults seldom exceed 1.5 meters and most are less than 1 meter. They are stouter than taipans but not thick-bodied like vipers. Unlike the mammal-eating taipans, tigersnakes eat mostly frogs.

The marine snakes of Asian and Australian waters are moderate-sized, mostly less than 1 meter as adults. Only a few species regularly exceed this adult size. The

Yellow-lipped seakrait
(*Laticauda colubrina*),
maximum known
length, 3.6 meters.

size champion among seasnakes is the yellow seasnake (*Hydrophis spiralis*); adults of this species have attained a reported maximum of 2.75 meters. The olive seasnake (*Aipysurus laevis*) has grown to 1.78 meters, but few adults exceed 1.1 meters. Similarly, Stoke's seasnakes (*Astrotia stokesii*), the heavyweights of marine snakes, are typically about 1 meter long as adults, although a few reach nearly 2 meters. Of the seakraits, adult female yellow-lipped seakraits (*Laticauda colubrina*) are often about 1 meter long and one individual grew to 3.6 meters.

.4.

SNAKEBITE

VENOMOUS OR POISONOUS: WHAT IS THE DIFFERENCE?

Poison is a broad term for any substance that irritates or kills. It is also used in a restricted sense for any harmful substance that enters the body by absorption through the skin or through eating or breathing. Poison ivy, for instance, irritates the skin; poison-dartfrogs kill predators that swallow them. Such plants and animals are called poisonous. *Venom* is a poison that one animal—whether a spider, a snake, or a bee—injects into another animal. Thus a snake or scorpion that injects a poison by biting or stinging is called venomous.

WHAT IS VENOM?

In snakes, venom is an evolutionary adaptation to immobilize prey, secondarily used in defense. Venoms are highly toxic secretions produced in special oral glands (see *How Do Fangs Work?*). Because these oral glands are related to the salivary glands of other vertebrates, venom can be considered a modified saliva. Venom immobilizes the prey when injected into its body, and in some cases initiates the digestive process by beginning the breakdown of the prey's tissues.

Each species has a unique venom with different components and different amounts of toxic and nontoxic compounds. The more closely related two species of snakes, the more similar their venoms. It is probable that venoms and venom

Western rattlesnake (*Crotalus viridus*)

As a timber rattlesnake (*Crotalus horridus*) swallows a rat, it alternately releases its jaws' grip on the left and right sides of the rat. Here, the right upper and lower jaws are open and shifting forward, exposing the right fang, encased in its sheath.

mechanisms evolved several times among snakes, increasing the diversity of venom chemistry and of the anatomy of the venom apparatus.

The chemistry of snake venoms is complicated. Venoms are at least 90% protein (by dry weight), and most of the proteins in venoms are enzymes. About 25 different enzymes have been isolated from snake venoms, 10 of which occur in the venoms of most snakes. Proteolytic enzymes, phospholipases, and hyaluronidases are the most common types. Proteolytic enzymes catalyze the breakdown of tissue proteins. Phospholipases, which occur in almost all snakes, vary from mildly toxic to highly destructive of musculature and nerves. The hyaluronidases dissolve intercellular materials and speed the spread of venom through the prey's tissue. Other enzymes include collagenases, which occur in the venom of vipers and pitvipers and promote the breakdown of a key structural component of connective tissues (the protein collagen); ribonucleases, deoxyribonucleases, nucleotidases, amino acid oxidases, lactate dehydrogenases, and acidic and basic phosphatases all disrupt normal cellular function, causing the collapse of cell metabolism, shock, and death.

Not all toxic chemical compounds in snake venoms are enzymes. Polypeptide toxins, glycoproteins, and low-molecular-weight compounds are also present in the venom of mambas and colubrids. The roles of most other components of venom are largely unknown.

Every snake's venom contains more than one toxin, and in combination the toxins have a more potent effect than the sum of their individual effects. In general, venoms are described as either *neurotoxic* (affecting the nervous system) or *hemotoxic* (affecting the circulatory system), although the venoms of many snakes contain both neurotoxic and hemotoxic components.

Venom components are broadly categorized by how they work to disrupt normal function.

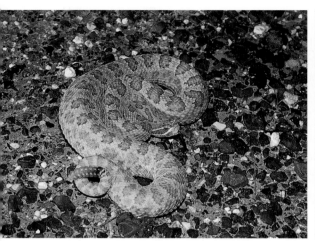

The western rattlesnake (*Crotalus viridis*; *C. v. viridis* [left], *C. v. helleri* [right]), contains nine geographic races differing in coloration and scalation. This rattlesnake also has several different venom races, but the distribution of the venom races does not match the distribution of the geographic races.

Enzymes—found in all snake venoms—initiate physiologically disruptive or destructive processes.

Proteolysins—found mostly in viper and pitviper venom—dissolve cells and tissue at the bite site, causing local pain and swelling.

Cardiotoxins—associated mostly with elapids and vipers—have variable effects; some depolarize cardiac muscles and alter heart contraction, causing heart failure.

Hemorrhagins—occurring in the venom of vipers, pitvipers, and the king cobra—destroy capillary walls, causing hemorrhages near and distant from the bite.

Coagulation-retarding compounds—found in some elapids—prevent blood clotting.

Thromboses—which some vipers have—coagulate blood and foster clot formation throughout the circulatory system.

Hemolysins—which are in the venom of elapids, vipers, and pitvipers—destroy red blood cells.

Cytolysins—components of viper and pitviper venom—destroy white blood cells.

Neurotoxins—found in elapids, vipers, tropical rattlesnakes, and some populations of the North American Mojave rattlesnakes (*Crotalus scutulatus*)—block the transmission of nerve impulses to muscles, especially those associated with the diaphragm and breathing.

Venom composition can vary among individuals of the same species, and even in the same litter, but variation is greater among geographically different populations.

For example, Mojave rattlesnakes from eastern Arizona and adjacent New Mexico have a special neurotoxin known as Mojave toxin, but their venom lacks hemorrhagic and some proteolytic properties. Venom from Mojave rattlesnakes of central Arizona lacks the Mojave toxin but has strong hemorrhagic and proteolytic properties. Where the two populations overlap, individual rattlesnakes have a venom with intermediate properties. Other vipers known to have venom races are the North American rock rattlesnakes (*Crotalus viridis*), North American copperheads (*Agkistrodon contortrix*), Malaysian moccasin (*Calloselasma rhodostoma*), and Russell's viper (*Daboia russelii*). Such venom variance is usually associated with the evolution of venom for capturing different prey. Because of these differences in venom composition, antivenom developed from and for one population often is not effective for bites from snakes of another population (see *What is Antivenom?*).

Venom toxicity may also vary over time in the same individual. Generally speaking, the venom of newborn and small juvenile snakes appears to be more potent than that of adults of the same species. Also, a bite from a snake that has not fed recently, such as one that has just emerged from hibernation, is more dangerous than the bite of one that has recently fed, because it has more venom to inject. Venom glands must replace venom lost with each strike-bite, and full replacement takes time.

HOW MANY SNAKES ARE VENOMOUS?

Four families of snakes (Atractaspididae, Colubridae, Elapidae, and Viperidae) include species dangerous to humans, a total of about 600 species or about 21% of all snake species. In none of these families are all species lethal to humans, although all atractaspidine, elapid, and viperid snakes are venomous. Generally speaking, the venoms most dangerous to humans are those of snakes that specialize in warm-blooded prey. Because human physiology is similar to that of their prey, the venoms react similarly in humans. But humans are also sensitive to snake venoms adapted to kill prey other than birds or mammals.

Danger may vary with the volume of venom injected. Even a mildly toxic venom is lethal if the snake injects enough of it. Conversely, a snake with a highly toxic venom is not dangerous if it is small and incapable of breaking the skin, or if in defense it does strike and bite but not inject venom. The Sonoran coralsnakes (*Micruroides euryxanthus*) have small mouths and usually do not break the skin when they bite. Some species have venom-delivery systems that do not permit them to deliver venom efficiently to large animals (see *How Do Fangs Work?*). Other species rarely come in contact with humans.

In the large family Colubridae, about one-quarter of the species (more than 600

Although most rear-fanged colubrid snakes have some toxic components in their saliva, few are rated as truly venomous. The boomslang (*Dispholidus typus*), in contrast, ranks within the top ten of venomous snakes.

species) have fangs, or at least enlarged and grooved maxillary teeth. But only four species have caused human fatalities: the African boomslang (*Dispholidus typus*), Oriental tigersnake (*Rhabdophis tigrinus*), Kirtland's birdsnake (*Thelotornis kirtlandii*), and Peruvian gray falseviper (*Tachymenis peruviana*). In the other three families, all species have fangs and should be considered potentially dangerous, although some are small or have venoms with weak effects on humans.

The salivas of some nonvenomous colubrids have, in rare instances, caused mild to moderate poisoning in humans. In the United States, people have had reactions to bites of the black-striped snake (*Coniophanes imperialis*), ringneck snake (*Diadophis punctatus*), western hognose snake (*Heterodon nasicus*), cat-eyed snake (*Leptodeira septentrionalis*), Mexican vinesnake (*Oxybelis aeneus*), western terrestrial gartersnake (*Thamnophis elegans*), common gartersnake (*T. sirtalis*), and lyre snake (*Trimorphodon biscutatus*). None of these snakes' venom-delivery systems operates efficiently on humans; to successful envenomate a human, the snake must be large and give a chewing bite for the venom to enter the wound. Symptoms of envenomation appear in fewer than 1% of gartersnake bites, although such bites are common among people who handle these snakes. It is possible that the salivas of all colubrids have a toxic component, and that some people are more susceptible than others.

WHERE ARE VENOMOUS SNAKES FOUND?

Venomous snakes inhabit all the continents except Antarctica (Figure 4.1). They also live on islands off the coasts of these continents and even on some remote oceanic islands. The snake fauna of each continent is different, and different venomous groups are dominant in each region (Table 4.1).

In the Americas, pitvipers (Crotalinae) predominate; many coralsnakes (Elapinae) are present in the tropical areas, but true vipers (Viperinae) are absent. The continental United States has 15 species of rattlesnakes (*Crotalus, Sistrurus*); 2 moccasins, the copperhead and cottonmouth (*Agkistrodon*); 2 coralsnakes (*Micruroides, Micrurus*); and the yellow-bellied seasnake (*Pelamis*, Hydrophiinae), a rare visitor to the southern California coast and the Hawaiian Islands. Venomous snakes have been recorded in every state except Alaska. In Canada, only southernmost Ontario, Manitoba, Saskatchewan, and British Columbia have a venomous snake.

The venomous snake fauna of Latin America is dominated by pitvipers and coralsnakes; a seasnake and two potentially dangerous colubrids, the Peruvian falseviper (*Tachymenis peruviana*) and Olfers' bushracer (*Philodryas olfersii*), are also present.

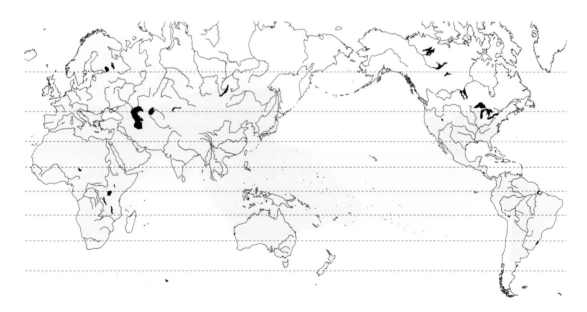

Figure 4.1. Distribution of venomous snakes throughout the world. No snakes live in the white areas. Green areas contain no venomous species of snakes. The yellow areas has one and only one species of venomous snake. The red areas contain two or more species of venomous snakes. Seasnakes and seakraits occur both in the open ocean and around islands, although most species live in relatively shallow water.

TABLE 4.1. OCCURRENCES OF VENOMOUS SNAKE SPECIES, BY MAJOR GEOGRAPHIC REGION

Family or subfamily	North America	Latin America[a]	Europe and the Middle East	Africa	Asia[b]	Australia and New Guinea	Pacific oceanic islands
Colubridae	0	2	0	2	1	0	0
Atractaspidinae	0	0	2	17	0	0	0
Hydrophiinae							
seakraits	0	0	0	0	3	2	4
seasnakes	0	1	1	1	40	32	12
terrestrial	0	0	0	0	0	100	3
Elapinae	2	64	1	24	46	0	0
Azemiopinae	0	0	0	0	1	0	0
Viperinae	0	0	21	46	18	0	0
Crotalinae	30	69	0	0	59	0	0
Total	32	135	25	89	167	134	11

Notes: Values denote approximate number of species. Some species occur in two or more regions.

[a]Central and South America, including the Caribbean islands.

[b]Asia, including Indonesia and the Philippines.

Europe and the Middle East have the fewest species of venomous snakes, most of which are true vipers. One species, the common European viper (*Vipera berus*), has a range extending from Scotland eastward all the way to the Pacific coast of Russia.

Africa's rich venomous snake fauna is dominated by true vipers (including arboreal vipers, desert vipers, nightadders, puffadders, and saw-scaled vipers), terrestrial elapids (cobras and mambas), molevipers (Atractaspidinae), and two of the world's four deadliest colubrids—the boomslang (*Dispholidus typus*) and the Cape birdsnake (*Thelotornis capensis*).

Asia is home to many pitvipers, including more than 30 species of *Trimeresurus*, abundant terrestrial elapids (cobras, the king cobra, kraits, and others), several marine elapids (seasnakes and seakraits), a few true vipers, and Fea's viper (Azemiopinae: *Azemiops feae*). One colubrid capable of fatal bites, the oriental tigersnake (*Rhabdophis tigrinus*), occurs in Japan and much of China. The red-necked keelback (*Rhabdophis subminiatus*) is potentially dangerous, but no human deaths have been reported for it.

The venomous snakes of Australia and New Guinea are the terrestrial elapids and marine seasnakes and seakraits. This fauna includes some of the deadliest snakes in the world. In southern Australia, the elapids constitute 80 to 100% of the terrestrial snake fauna. Most snakes encountered in Australia are likely venomous.

The waters of the South Pacific are home to a few seasnakes and seakraits, particularly the tropical waters of Australia and the East Indies. A small terrestrial elapid, the bola (*Ogmodon vitianus*), is found on Fiji. Several other elapids occur on the Solomon Islands, including the jungle shark (*Loveridgelaps elapsoides*).

WHICH SNAKES HAVE THE MOST POTENT VENOM?

This question can be answered in two ways, yielding different answers. One way estimates lethality based on the potential amount of venom that a snake might deliver with a single bite. However, no snake empties its venom glands with a single bite, and occasionally a snake delivers a "dry" bite. The other way of estimating lethality is to test a venom's killing power on mice.

The mouse-test procedure estimates the strength of a venom by injecting measured amounts into a large sample of mice and recording the dosage of venom that kills 50% of the mice within 24 hours. This dosage, called the LD_{50} (*LD* standing for "lethal dose"), is measured in milligrams of venom per kilogram of mouse. The venoms of many species of snakes have been characterized in this way. For obvious reasons, the lethal doses for humans have not been determined.

Small or large, Russell's vipers (*Daboia russelii*) are highly dangerous, and their bites result in one of the highest mortality rates in Asia. Because different populations have different venom compositions and comprise different venom races, antivenom produced in one country is not necessarily effective in an adjacent country.

The LD_{50} values for certain species are listed in Table 4.2. Assuming that mice and humans have similar susceptibilities to the venom of each species, that of the hook-nosed seasnake (*Enhydrina schistosa*) is the most lethal venom tested so far; it is estimated that only 1.5 milligrams of its venom will kill a human being. The hook-nosed seasnake ranges from the Persian Gulf and the waters of southern Asia to the northern coast of Australia. Russell's viper (*Daboia russelii*) of southern Asia and the inland taipan (*Oxyuranus microlepidotus*) of Australia are nearly as deadly. Of course, these comparisons are only estimates of the venom's toxicity in humans. Further, these LD_{50} values are mixed data, derived from different studies using different sites of venom injection (intramuscular, intraperitoneal, and subcutaneous). Subcutaneous injections are typically less lethal than intraperitoneal ones, and may require 2 to 5 times the venom dosage to obtain the same kill rate.

Estimates of yield, such as the ones listed in Table 4.2, require forcibly draining (milking) all venom from a snake's venom glands, drying the sample, and then weighing the powdery residue. This milking procedure is performed on a sample of adult snakes of a single species, and the average yield is determined. King cobras (*Ophiophagus hannah*), Gaboon vipers (*Bitis gabonica*), eastern diamondback rattlesnakes (*Crotalus adamanteus*), and bushmasters (*Lachesis*) have the capacity to deliver the largest volume of venom in a single bite. The toxicity of their venoms differs, but the bites of all four are highly dangerous even if they inject only one-quarter of their venom supply. Approximately 100 milligrams of venom from an eastern diamondback rattlesnake kills an adult man, and large diamondbacks store as much as 850 milligrams in their venom glands!

The deadliness of a venom varies with the prey. Even if we confine our attention to humans—although humans are not the prey of any snake—many factors b̲e̲s̲ides yield and LD_{50} values influence the seriousness of a bite. In humans, the fac̲t̲ in̲-clude the individual's health, size, age, and psychological state. Factors ̲ ̲ted with the nature of the bite include penetration of one or both fangs, a̲ ̲ ̲t of venom injected, location of the bite, and proximity to major blood vessels. The health of the snake and the interval since it last used its venom mechanism also enter in. These variables make every bite unique. Depending on circumstances, the bite of a "mildly" venomous snake may be life-threatening and that of a "strongly" venomous snake may not.

HOW DO FANGS WORK?

A fang is simply a tooth modified to inject venom into prey. The fangs work in concert with other structures to form a complete venom-delivery apparatus that functions like a hypodermic syringe and needle (Figure 4.2).

TABLE 4.2. TOXICITY OF SELECT SNAKE VENOMS

	Mouse LD$_{50}$ (mg/kg)[a]	Venom yield per snake (mg)[b]
Hook-nosed seasnake (*Enhydrina schistosa*)	0.02	7.0–79.0
Russell's viper (*Daboia russelii*)	0.03	130.0–250.0
Inland taipan (*Oxyuranus microlepidotus*)	0.03	44.0–110.0
Dubois's seasnake (*Aipysurus duboisii*)	0.04	0.4–0.7
Eastern brownsnake (*Pseudonaja textilis*)	0.05	2.0–67.0
Black mamba (*Dendroaspis polylepis*)	0.05	50.0–100.0
Tiger rattlesnake (*Crotalus tigris*)	0.06	6.0–11.0
Boomslang (*Dispholidus typus*)	0.07	1.6–8.0
Yellow-bellied seasnake (*Pelamis platurus*)	0.07	1.0–4.0
Common Indian krait (*Bungarus caeruleus*)	0.09	8.0–20.0
Saharan horned viper (*Cerastes cerastes*)	0.10	20.0–45.0
Common taipan (*Oxyuranus scutellatus*)	0.10	120.0–400.0
Common European viper (*Vipera berus*)	0.11	10.0–18.0
Tigersnake (*Notechis scutatus*)	0.12	35.0–189.0
Forest cobra (*Naja melanoleuca*)	0.12	?
Puffadder (*Bitis arietans*)	0.14	100.0–350.0
Gaboon viper (*Bitis gabonica*)	0.14	350.0–600.0
Brown-lipped seakrait (*Laticauda laticaudata*)	0.16	?
Neotropical rattlesnake (*Crotalus durissus*)	0.17	20.0–100.0
Mojave rattlesnake (*Crotalus scutulatus*)	0.18	50.0–150.0
Egyptian cobra (*Naja haje*)	0.19	175.0–300.0
Harlequin coralsnake (*Micrurus fulvius*)	0.20	3.0–5.0
Kirtland's birdsnake (*Thelotornis kirtlandii*)	0.21	?
Saw-scaled viper (*Echis carinatus*)	0.24	5.0–48.0
Eastern green mamba (*Dendroaspis angusticeps*)	0.26	60.0–95.0
Spectacled cobra (*Naja naja*)	0.28	150.0–600.0
Common lancehead (*Bothrops atrox*)	0.35	100.0–200.0
Common deathadder (*Acanthophis antarcticus*)	0.60	70.0–236.0
King cobra (*Ophiophagus hannah*)	0.90	350.0–500.0
Western rattlesnake (*Crotalus viridis*)	1.01	35.0–250.0
Banded krait (*Bungarus fasciatus*)	1.20	20.0–114.0
Malayan pitviper (*Calloselasma rhodostoma*)	1.24	40.0–60.0
Timber rattlesnake (*Crotalus horridus*)	1.64	75.0–210.0
Cottonmouth (*Agkistrodon piscivorus*)	2.04	80.0–170.0
Western diamondback rattlesnake (*Crotalus atrox*)	2.20	175.0–600.0
Sidewinder (*Crotalus cerastes*)	2.60	18.0–50.0
Pygmy rattlesnake (*Sistrurus miliarius*)	2.80	12.0–35.0
Massasauga (*Sistrurus catenatus*)	2.90	15.0–45.0
South American bushmaster (*Lachesis muta*)	4.50	200.0–500.0
American copperhead (*Agkistrodon contortrix*)	10.90	40.0–75.0

Notes: Species are listed approximately in order from most to least toxic, in terms of the dose of venom that kills 50% of the test mice within 24 hours of injection (LD$_{50}$). Missing values are unknown.

[a]LD$_{50}$ values given here are not strictly comparable to one another because they may have been based on different injection sites.

[b]Venom yield is given as the dry weight of all the venom that could be milked from an adult snake.

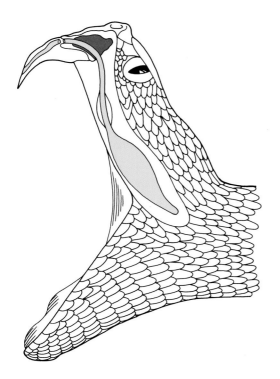

Figure 4.2. The venom apparatus of a viper. The venom gland is encased in a sheath of muscles (not shown) above the angle of the jaws. Venom flows through the venom duct to the base of the fang, into the fang's venom canal, and out into the prey. (Adapted from Carr 1963)

Venom is produced by a pair of large venom glands. One gland is located on each side of the head, below and behind the eye, above the upper rear corner of the jaw. In some species, the gland extends backward along the neck, and in African night-adders (*Causus*) the gland can extend to midbody and even beyond. Within these glands, which are typically almond- or pear-shaped, the venom is produced by several (usually 4 to 5) lobes of secretory cells. The secretory cells can make up as much as 80% of the gland's total cell content. Their secretions drain through small tubules into a hollow space, the lumen of the gland. The lumen in turn joins the venom duct, which carries the venom forward to the base of the fang. To continue the syringe analogy, the venom gland corresponds to the body of the syringe, and the venom duct to its throat. The venom duct is surrounded by small masses of glandular tissue, the accessory glands, which may act as valves to regulate the flow of venom to the fang. Even though the accessory glands' secretions are not toxic, they may activate some venom components: venom drawn from the lumen of the venom gland is less toxic than venom taken from the fang.

The venom duct does not extend into the fang. It opens adjacent to the fang, within a sheath of connective tissue surrounding the fang's base. This sheath is a seal around the fang, directing the flow of venom into the fang's canal and outward into the prey.

Fangs are wide at their bases and gradually taper to needlelike points. All snakes' fangs are curved. The amount of curvature varies among species; in rattlesnakes (*Crotalus*), for instance, the middle of each fang forms an arc of 60 to 70 degrees. The broad base of each fang sits in a socket of the maxillary bone and contains an opening adjacent to the end of the venom duct. The venom flows into the fang's venom canal, which extends downward through the fang to a discharge orifice on the front surface of the fang, just above its solid tip. The discharge orifice is an elongated slit whose size varies among species.

Both the venom canal and the outer surface of the fang are covered with enamel. The presence of enamel on both surfaces is a clue to the evolution of fangs. The first change appears to have been the appearance of a groove on the outer surface of one or more maxillary teeth. Concurrent with selection for more effective venom, the grooved teeth then enlarged and the grooves deepened. Eventually the walls of the groove closed over it, forming a closed canal. Among the evidence for this hypothesis of fang evolution is a faint seam on the face of each fang in some elapid and viperid snakes, indicating the point of contact and fusion of the two sides. Other evidence includes partially closed venom canals, such as in the African nightadders (*Causus*).

How Is Venom Injected?

A snake's fang corresponds to a syringe's needle. In fact, the free end of the fang is identical to the tip of a needle. Both have sharp tips to penetrate skin and muscle, and discharge orifices near the tip. Finally, the jaw musculature surrounding the venom gland corresponds to the syringe's plunger (Figure 4.2). The contraction of these muscles squeezes the gland, forcing venom from the lumen into the venom duct and outward through the fang.

Among the various venomous snakes, biologists have identified three distinct venom-delivery systems. There is evidence that each system evolved more than once. For instance, the folding fang of vipers and pitvipers is also found in the Australian deathadders and African stilettovipers, three groups that are not closely related. All three systems evolved from the basic snake tooth, which is slightly curved and cone-shaped. This basic *aglyphous* (grooveless) tooth (*a*, without; *glyphe*, carving or groove) occurs in all snakes, even those with fangs, and most snakes have only these ungrooved teeth. Evolution of grooved and, eventually, canaled teeth (fangs) occurred only on the maxillary bone of the upper jaw. The three types of venom-delivery systems differ with regard to the position of the fang on the maxillary, the nature of the venom groove or canal, and the mobility of the fang-maxillary unit.

Ancestral snakes were venomless and had only grooveless teeth. All blindsnakes, boas, pythons, and other henophidian snakes still have exclusively aglyphous teeth

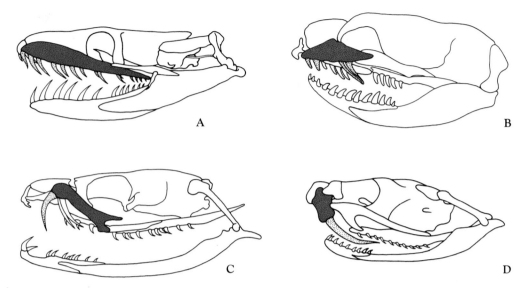

Figure 4.3. The four general types of maxillary dentition of snakes. The maxillary bone is stippled in each figure. (A) Aglyphous snakes have no grooves or canals on their teeth; skull of a Central African rock python, *Python sebae*. (B) Opisthoglyphous snakes have a groove on one or more teeth on the rear of the maxillary bone (red); skull of a robust glossysnake, *Polemon robustus*. (C) Proteroglyphous snakes have a fang on the front edge of each largely immobile maxillary bone; skull of a Jameson's green mamba, *Dendroaspis jamesoni*. (D) Solenoglyphous snakes have a hinged maxillary bone, allowing the large fangs to be erected as the mouth opens; skull of a timber rattlesnake, *Crotalus horridus*.

(Figure 4.3A). Most colubrid snakes—the largest and most diverse snake family—are also aglyphous, although some species have 1 or 2 enlarged rear teeth on each maxillary bone. These enlarged teeth may be separated from the front maxillary teeth by a gap known as a *diastema*. A diastema commonly separates the enlarged rear teeth, whether grooved or not, from the smaller front maxillary teeth.

Snakes with enlarged rear maxillary teeth are termed *opisthoglyphous* (*opistho*, behind) (Figure 4.3B). In some opisthoglyphous colubrids the enlarged teeth are ungrooved, but most others have a groove on the face or the side of the enlarged tooth. Originally these elongated teeth probably served only to hold prey, and this use persists in gartersnakes (*Thamnophis*). However, such teeth puncture the prey's skin, and some saliva inevitably enters the puncture wounds. An adaptive advantage would have resulted from increasing the toxicity, digestive ability, or tranquilizing effects of saliva, and assuring its delivery deep into the wound. Any of these advantages would have driven the evolution of fangs and venom glands, producing a range of venoms and venom-delivery systems.

Evolution of venom-delivery systems proceeded in two main directions: toward fixed *proteroglyphous* fangs (*protero*, earlier) and toward hinged *solenoglyphous* fangs (*soleno*, pipe). In both instances, the grooved tooth became a fang by virtue of closure of the groove and a shifting forward of the enlarged tooth to the front of the mouth.

In proteroglyphous snakes, the fangs are short because large fixed fangs would require a deepening of the mouth cavity to prevent the fangs from perforating the floor of the mouth. The proteroglyphous condition (Figure 4.3C) is typical of the elapid snakes (cobras, taipan, coralsnakes, seasnakes and their relatives). In many species, particularly the seasnakes, the fang is barely longer than the teeth behind it. Proteroglyphous snakes typically bite and hold their prey, and then chew to inject venom deep in the wound. This behavior is virtually universal among seasnakes, whose fish prey would otherwise swim or drift away before being incapacitated by the venom. Probably only the largest proteroglyphous snakes can deliver an envenomating strike-bite and withdraw. The largest elapid, the king cobra (*Ophiophagus hannah*), has fangs only 8 to 10 millimeters long; fangs are less than 8 millimeters long in mambas (*Dendroaspis*), less than 7 millimeters in Indian cobras (*Naja naja*), and less than 3 millimeters in adult harlequin coralsnakes (*Micrurus fulvius*) and yellow-bellied seasnakes (*Pelamis platurus*).

The hinged fangs (Figures 4.3D and 4.4) of the vipers and pitvipers (Viperidae) represent a more intricate system that allows a snake to strike, envenomate, and withdraw from the struggling prey, thereby avoiding injury. The hinged fang sits at the front of the mouth on a short maxillary bone that can rotate forward and backward. When not in use, the fang folds backward and upward against the roof of the mouth, where it lies enclosed in a membranous sheath. During a strike, the maxilla rotates forward, erecting the fang, and the mouth opens nearly 180 degrees (Figure 4.4). As the mouth strikes the prey, the jaws close, propelling the fangs into the prey; the venom is injected at the time of penetration. The right and left fangs can be rotated independently, although they erect jointly. A viper often works its fangs back into their resting sheaths one at a time after swallowing its prey.

The advantage of folding fangs is that long fangs can be housed in the mouth without perforating the floor of the mouth. Viperids have significantly longer fangs than the proteroglyphous elapids, and some viperids seem to have taken the evolutionary opportunity of lengthening their fangs to the extreme. *Bitis*, a group of African vipers, have the longest fangs known: up to 28 millimeters in the puffadder (*B. arietans*), and more than 30 millimeters in large Gaboon vipers (*B. gabonica*). Even in the smaller copperhead (*Agkistrodon contortrix*) and common European viper (*Vipera berus*), fangs are 7 millimeters or longer.

Folding fangs occur in two other groups of snakes. The Australian deathadders (*Acanthophis*), although they are elapids, are solenoglyphous. Their folding-fang mechanism is very similar in appearance and operation to that of the vipers and

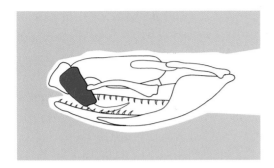

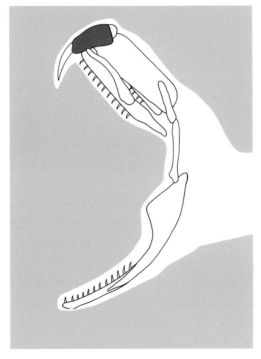

Figure 4.4. Operation of the hinge mechanism of a solenoglyphous skull: (top) the fang's position when the mouth is closed; (bottom) the erect fang position immediately prior to biting the prey (also see Figure 4.3D). (Adapted in part from Kardong 1974)

pitvipers. The deathadders also have the body shape and ambush-hunting habits of many viperids, an excellent example of convergent evolution.

The African molevipers (*Atractaspis*, Atractaspidinae) are also solenoglyphous. Their short maxillary bones rotate and bear long fangs, but their strike-bite differs from that of the viperids and deathadders. They are burrowers, and the confines of narrow burrows make a typical rearing strike impossible. Instead they crawl alongside their prey, open their mouths slightly, and shift the lower jaw away from the prey, freeing the fang nearest the prey. With a backward and sideward stab, they embed the fang and inject venom into their prey (typically newborn rodents and burrowing lizards). Because they stab backward, rather than biting forward, a snake handler who grabs one behind the head often ends up with a fang embedded in a finger or thumb; this accounts for the snake's common name, stilettoviper.

Newborn venomous snakes are fully operational. They have fangs and inject venom when they bite. Throughout the lives of all snakes, however, teeth and fangs are shed and replaced regularly. An ordinary tooth is replaced by one that forms beneath it, eventually loosening and then pushing it out of its socket. Proteroglyphous and solenoglyphous fangs are replaced in a somewhat different fashion. A series of 5 to 7 replacement fangs lies in the gums behind and above the functional fang. These replacement fangs are arranged in a graduated series, the largest adjacent to the functional fang. As the functional fang wears down, it is replaced by the next fang. The reserve-fang series then shifts forward, so that a replacement fang is always available to replace a damaged functional fang.

The replacement fangs do not develop fully formed but in miniature; instead, the growth process forms the tip first and then builds up the base, thus enlarging the fang and pushing the tip outward. The hollow-needle shape is apparent early in development. Functional fangs are shed in cycles as short as 10 days and as long as 6 to 10 weeks, depending on the species and the health of the individual snake. During the replacement phase, a snake may briefly have two fangs on each side of its head.

Do Snakes Spit Their Venom?

Some cobras can spray their venom up to 2.5 meters. This action is called spitting, but it does not involve puckering the lips and blowing the venom outward. Spitting is a defensive behavior that has nothing to do with killing prey. Spitting cobras bite and envenomate their prey just as do other venomous snakes.

Venom-spitting apparently evolved in three separate cobra groups (family Elapidae) but in no other snake families. Two of the spitting-cobra groups are African; one group is the African ringhal cobra (*Hemachatus haemachatus*); the other group includes the black-necked cobra (*Naja nigricollis*), the Mozambique spitting cobra (*N. mossambica*), the Mozambique red spitting cobra (*N. pallida*), and the West African spitting cobra (*N. katiensis*). The third group of spitters occurs in southern Asia and includes the golden spitting cobra (*Naja sumatrana*) of the Malay Peninsula and Sumatra, the Indonesian spitting cobra (*N. sputatrix*) of southern Indonesia, the common spitting cobra (*N. philippinensis*) and Samar spitting cobra (*N. samarensis*) of the Philippines, the Burmese spitting cobra (*N. mandalayensis*), the Chinese and Indochinese populations of the Asian black cobra (*N. atra*), and some populations of the widespread Asian monocled cobra (*N. kaouthia*). These snakes live in areas inhabited by large herbivores that might trample them or large carnivores that might eat them, so they spit their venom defensively.

Spitting or spraying of venom involves no major evolutionary structural modification. The fangs of spitting cobras resemble those of their nonspitting relatives, except that the discharge orifice of the fang is greatly reduced in size and pointed

Ringhals (*Hemachatus haemachatus*) are one of at least two spitting cobras in Africa. Spitting cobras are dangerous, both by their defensive spraying of venom into the eyes of a disturber and a highly toxic venom from a bite.

more forward. When compression of the venom gland forces its secretion through the venom duct and hollow fang, the venom is not discharged from the fang as quickly as in a snake with a normal-sized discharge orifice. The venom thus backs up in the fang, creating greater pressure at the discharge opening than in a normal fang, and the venom sprays from the fang in tiny droplets instead of large drops. The snake aids expulsion of venom by forcibly collapsing its lung and blowing air out of its mouth. The air carries the venom in a pair of fine sprays aimed at the eyes of the intruder.

At close quarters, the spitting cobras have very accurate aim. If the neurotoxic venom reaches the eyes, it is quickly absorbed by the capillaries of the conjunctiva. The venom may cause temporary blindness by irritating the cornea; extensive damage of the cornea can lead to permanent blindness. The venom should be rinsed out of the eye as soon as possible.

Reports occasionally surface of venom-spitting by vipers or pitvipers (Viperidae), some of which may sling venom around if agitated and striking violently. Some small West Indian boas of the genus *Tropidophis* are said to spit blood when disturbed; these reports may be faulty observations of these snakes' defensive behavior of dripping blood from their eyes. The only true venom-spitting snakes are cobras.

How Dangerous Is a Snakebite?

A snakebite is usually not dangerous, unless it involves one of the more than 200 species that produce a potent venom. Every day, people are bitten by nonvenomous snakes and experience only the slight discomfort caused by the snake's teeth puncturing or scratching the skin. Of course, such wounds may be painful if the snake has long teeth, such as those of a python or large ratsnake, but serious effects are rare. Bites by nonvenomous species can be treated by washing the wound and applying an antiseptic to the punctures or scratches. Bites by venomous snakes require medical treatment (see *If Bitten, What Next?*). If untreated, a venomous bite may result in serious tissue or organ damage and even death. Serious secondary bacterial infections, such as gas gangrene and tetanus, may also follow venomous snakebites, and loss of a limb, finger, or toe is not uncommon. A nonvenomous snakebite usually involves several puncture marks of equal depth; that of a venomous snake is characterized by one or two larger and deeper punctures among more shallow marks. However, tooth marks are not a reliable method for identifying the potential danger of a snakebite.

If the snake is venomous, discomfort is usually felt within a few minutes. A burning sensation or pulsating pain is often accompanied by swelling or discoloration of the tissues surrounding the wound. Such localized discomfort is particularly characteristic of hemotoxic envenomation by pitvipers and true vipers, but moderate to severe local pain may also accompany neurotoxic bites of some elapids.

Medical treatment should be obtained for all elapid bites, even when there is no pain. Serious elapid bites are not usually apparent, since immediate pain does not always occur. A characteristic early sign of a serious neurotoxic elapid bite is drooping eyelids, followed by difficulty in swallowing, slurred speech, severe thirst, vertigo, and difficulty in breathing. Later, blood pressure often drops, and cardiac arrest may occur. A full survey of the symptoms of venomous snakebites is beyond the scope of this book; more detail can be found in the further readings we have suggested for this topic.

The most comprehensive study, published a half century ago (Swaroop and Grab 1954), estimated that 300,000 venomous snakebites occurred throughout the world each year and 30,000 to 40,000 of these bites resulted in death. A more recent worldwide survey is not available, and it is likely that the 1954 figures were too low because data from China, the then Soviet Union, and central European nations were unavailable; also the number of snakebites from Africa appears to have been

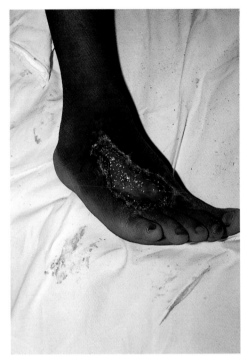

A nasty wound from the bite of a Central American bushmaster (*Lachesis stenophrys*). Prompt treatment of this farmer's snakebite allowed a quick recovery. (S. Hendrix)

an underestimate. It is probable that worldwide now more than 50,000 snakebite deaths occur each year (Warrell 1997). Death from snakebite is highest in developing nations with a large rural human population engaged in hands-on agriculture and with little access to medical treatment, and lowest in developed nations with many medical facilities.

The more rural the habitat, the greater the chance of encountering a venomous snake. The highest incidence of snakebite mortality occurs in Sri Lanka and India. In Sri Lanka, 5.7 individuals per 100,000 people die annually from snakebite. Nearly 70% of the bites occur to males during the harvesting season, according to de Silva and Ranasinghe (1983). Most of the reported bites are by the small hump-nosed viper (*Hypnale hypnale*) and occur most frequently on the feet and ankles; fortunately, less than 3% of this viper's bites result in death. In Sri Lanka and elsewhere on the Indian subcontinent, most deaths result from bites by the various kraits (*Bungarus*), cobras (*Naja, Ophiophagus*), saw-scaled vipers (*Echis*), and the Russell's viper (*Daboia russelii*). Mid-twentieth century estimates suggest that 7,000 to 15,000 people died annually of snakebite in the former British India; this estimate yields 4 to 5 deaths per 100,000 people, and recent data suggest that this mortality rate is probably about the same now. In the United States, by contrast, only about 10 to 15 people die each year from venomous snakebites and about half of these are from foreign species held as "pets." These data yield a fatality rate of fewer than 0.2 people per 10,000, a rate much less than deaths by stings of bees and wasps. Of the fatalities by native snakes, most were by the cottonmouth (*Agkistrodon piscivorus*), western rattlesnake (*Crotalus viridis*), and eastern and western diamondback rattlesnakes (*C. adamanteus, C. atrox*). In Europe the death rate from all venomous animal bites was less than 0.5 per 100,000 people. Today, snakebite mortality worldwide is probably about 50% of what it was when the preceding data were compiled. Modern medical treatment has improved survivorship, and treatment is more widely available.

How Do You Avoid Snakebites?

Do not handle venomous snakes. In Europe and North America, most snakebites occur when the victim is either holding a snake or attempting to pick up or kill it.

Never play with venomous snakes. Remain at a safe distance—no nearer than two snake body lengths—from the snake.

Do not pick up a "dead" snake! It may be only injured, stunned, or playing dead. Even with a truly dead snake, reflex action can cause the jaws to open and close. A fatal envenomation from the decapitated head of a canebrake rattlesnake (*Crotalus horridus atricaudatus*) has been reported (see *Do Injured Snakes Die before Sundown?*).

Bites from unseen snakes in the wild may be prevented by common sense and proper dress. Boots and coarse long trousers should be worn in such areas. Most bites that occur in the wild are on the extremities. Do not put your hands or feet in places that you have not visually examined first. At night, one's path should always be lighted to make snakes visible, since many venomous snakes are nocturnal.

If Bitten, What Next?

A person bitten by a venomous snake should be taken to a hospital immediately. The traditional cut-and-suck first-aid methods for snakebite are now subject to serious doubt. Because they involve cutting and constriction of blood flow, they can do more harm than good. Self-treatment is likely to worsen an already serious condition.

Before arriving at the hospital, (1) keep the patient as calm and still as possible; (2) immobilize the bitten limb, using a splint if possible and positioning it below the level of the heart; (3) do not perform such traditional measures as cooling with ice, applying a tourniquet, cutting and sucking, giving alcohol or aspirin, pouring turpentine onto the wound, and the like; and (4) if the bite is that of a neurotoxic snake, wrap the limb in a pressure bandage to localize the venom (a measure that has proven effective for bites of Australian elapids). Whenever possible, the snake responsible for the bite should be brought to the medical facility for purposes of identification. It is better to avoid a second bite, however, if the snake is difficult to capture. And neither capture nor first-aid measures should delay transport of the patient to a hospital.

At the hospital, encourage the medical staff to call a poison-control center for expert advice on snakebite treatment. Because snakebites are uncommon in the United States, few medical personnel have experience in treating them. Generally speaking, the recommended course of action is to observe the patient to determine the extent of envenomation. Venomous snakes can strike and bite entirely in defense without injecting venom. Such "dry" bites account for 20 to 40% of all snakebites.

The patient is typically observed for at least 8 hours, because the onset of some symptoms (particularly those of neurotoxic venom) may occur hours after the bite. Like all puncture wounds, bites must be thoroughly cleaned, and antitetanus serum and a broad-spectrum antibiotic are often recommended.

Antivenom is the only specific treatment for envenomation, and it should be given only to persons with symptoms or signs of envenomation. Antivenom should be administered only in a medical facility and only by a health-care specialist. The patient must first be tested for hypersensitivity to horse serum, if the antivenom derives from purified horse blood (see *What Is Antivenom?*). In the United States, only one anti-crotalid antivenom is now available for patients bitten by cottonmouths, copperheads, or rattlesnakes. This antivenom (CroFab™ produced from sheep serum) is used only for patients exhibiting symptoms of envenomation.

Antivenoms are developed for specific venomous snakes. Thus a European viperid antivenom would not neutralize the toxin of North American pitvipers because it was developed for different vipers.

WHAT IS ANTIVENOM?

Antivenom is a serum that is commercially produced to neutralize the effects of envenomation by venomous snakes. The fresh snake venom used to produce antivenom is obtained either by manually milking a snake or by electrical stimulation. Venom is extracted from captive snakes every 20 to 30 days. In manual milking, the snake is held behind its head and induced to bite a thin rubber diaphragm covering a collecting vessel while the handler applies pressure to the snake's venom glands. The pressure is maintained until no more venom is discharged. In electrical stimulation, electrodes are touched to the opposite sides of the snake's head, causing the muscles around the venom gland to contract, expelling venom into a collection container. The venom is freeze-dried (the preferred method), or dried with the help of a drying agent or a vacuum.

A white-lipped treeviper (*Trimeresurus albolabris*) swallowing a treefrog. A drop of venom is suspended on the tip of the left fang, likely extruded by pressure from the struggling frog. (R. W. VanDevender)

Healthy horses or sheep are injected at regular intervals with nonlethal doses of a saline solution prepared from the freeze-dried venoms, usually of several species, until they build up an immunity to the venom. The dosage can then be slowly increased over time to create greater immunity. The sheep's or horse's immune system neutralizes the venom by producing antibodies (specialized proteins). When injected into humans, these antibodies in turn neutralize the same venom.

To obtain the antibodies, a small amount of blood is regularly withdrawn from the immunized animal. The blood is combined with a sodium-citrate solution to prevent coagulation and degradation, and the globulin to which the antibodies are attached is separated out and purified. About 25 laboratories throughout the world produce antivenoms for the venomous snakes in their particular regions. Because venomous snakes are often genetically quite distinct in different countries, the antivenoms produced in one country, even an adjacent one, may be ineffective for the bites of the same or related species elsewhere. For example, the cobra antivenom produced in Thailand will not successfully counteract the venom of an Indian cobra.

CAN VENOM BE USED AS MEDICINE?

Snake venoms have great potential for medical use because of the wide variety of compounds they contain and the specific action of each compound. Although no medical preparation derived directly from snake venom is used now in the United

States, a few such compounds are used in Asia, Europe, and Latin America for treatment of blood disorders. Nowhere is a whole venom used as a medicine; instead, specific components are extracted.

Beta blockers—drugs widely used in the treatment of cardiovascular diseases—owe their discovery to research on *Bothrops* venoms. These venoms contain a peptide that interrupts the activity of an enzyme involved in hypertension (high blood pressure). Two analgesics derive from cobra venom: Cobroxin™ is used like morphine to block nerve transmission, and Nyloxin™ reduces severe arthritis pain. Arvin™, an extract of the Malayan pitviper (*Calloselasma*), is an effective anticoagulant (it inhibits the formation of blood clots).

Venom components are also used in basic research in physiology, biochemistry, and immunology. By retarding or speeding up biochemical and cellular processes, such as nerve-impulse transfer and blood clotting, venom components allow researchers to examine the operation of the process and to develop drugs to correct malfunctions due to disease. For example, venoms are currently being investigated for their potential as antiviral and antibacterial agents (tetanus, hepatitis, trachoma, scarlet fever, malaria, botulism) and anticarcinogenic agents (cancer and nonmalignant tumors). Other diseases for which snake venoms have been used in research include nerve diseases, such as epilepsy, multiple sclerosis, myasthenia gravis (Lou Gehrig's disease), Parkinson's disease, and poliomyelitis; musculoskeletal diseases, including arthritis and rheumatism; cardiovascular diseases, such as hypotension, hypertension, angina, and cardiac arrhythmias, and visual disorders, including neuritis, conjunctivitis, and cataracts (see *Why Are Snakes Important to Us?*).

.5.

SNAKES AND US

WHY ARE SNAKES IMPORTANT TO US?

Should the survival of a plant or animal species depend on its importance to humans? Equating the survival of a species with its importance to humans (which usually means its economic value) ignores humanity's responsibility for stewardship of other living things. Stewardship is not about improving the lives of other organisms, it involves ensuring that our activities do not diminish the ability of other species to survive (and, in doing so, ensuring our own survival as a species). Stewardship extends from stopping the car to remove a turtle or snake from the highway to establishing large nature reserves and reducing pollution. Our activities and our growing population worldwide are straining the fabric of the natural world. More and more species are threatened with extinction.

The earth has experienced mass extinction events several times in the distant past. One or more of these events occurred when the earth was struck by large meteorites. The causes of the other mass extinctions, in which more than 30% of the earth's plant and animal species disappeared, are unknown. What is certain is that the community of life changed dramatically after each extinction event. And there is no assurance that humans can survive a mass extinction event, even if we are the ones to trigger it. Extinction is forever!

Although our knowledge of ecology is vast, we do not yet know how many and which species are necessary for the operation of an ecological community. We use terms such as *keystone species* to signify that the loss of certain organisms will cause

Emerald treeboa (*Corallus caninus*)

129

The leopard ratsnake (*Elaphe situla*) is swallowing a rat. Snakes in the genera *Elaphe* and *Ptyas* received their common name ratsnake because these rodents are a principal food of most species in these two groups.

the demise of a community. It is questionable whether any snake is a keystone species, but this does not lessen their individual importance to natural and human communities.

All snakes are predators. Many species specialize in hunting rodents, a dietary preference that should, from a human perspective, place a premium on snakes' survival. Indeed, many farmers in the United States and abroad protect snakes because they keep rats and mice in check, reducing losses of stored grains and other food-stuffs. An adult 1-meter ratsnake (*Elaphe*) will eat an average of one adult rat a week. A ratsnake can go everywhere a rat goes and is exceptionally good at locating litters of newborn rats.

Are the benefits of rodent-eating snakes apparent on a large scale? In India, where snakes are harvested heavily for the leather trade, rodent outbreaks have occurred; local Indian governments now recognize the value of snakes and encourage their protection. During the DDT-induced collapse of hawk populations in coastal

Bottling a venomous snake in an alcoholic drink is a common folk medicine practice in many Asian nations. Cobra wine, a local product sold in Hanoi's Le Mat snake village, contains the local monocled cobra (*Naja kaouthia*). (M. Russell)

areas of Florida in the 1960s, rattlesnake populations expanded to feed on rodents no longer subject to hawk predation. In the southwestern United States, rodents and their fleas carry the plague bacillus. If rodent control by nonvenomous (*Coluber, Elaphe, Masticophis, Pituophis*) and venomous (*Crotalus*) snakes is reduced, human exposure to rats and hence to the plague bacillus rises. Recently a new threat, the hantavirus, has appeared in the Southwest and in Florida. That rodents are the prime vector for this virus represents another reason to protect snakes for purposes of rodent control.

Commercially, snakeskin remains a minor but not insignificant aspect of the leather trade. Beginning in the 1980s, snakes and other reptiles became a multi-million-dollar pet-trade industry. Thousands of snakes are imported into North America, Europe, and Japan each year for snake fanciers, and captive rearing of snakes for sale (herpetoculture) continues to grow. As pets, snakes support a subindustry of pet suppliers, including rodent farming for snake food. Snakes are a common food item in the markets and eateries of eastern Asia, and are not unknown as food in North America and Europe.

Snakes and snake parts are used widely in the folk medicines of Asia. The effectiveness of snake parts for curing various illnesses is highly questionable, as is their use as aphrodisiacs or virility boosters. Liquor companies in Asia raise snakes for insertion in bottles of high-priced whiskey. (One such company in Japan also supports a research laboratory of snake biology.)

Snake venoms are widely used in biomedical research. Venoms consist of numerous components (see *What Is Venom?*) that alter blood composition, circulatory control, and nerve transmission. These compounds are used in experiments to examine the cellular and subcellular mechanisms associated with nerve-impulse transfer, pain reduction, clotting, reduction of blood pressure, and other processes (see *Can Venom Be Used as Medicine?*). To date, only a few drugs derive from purified venom, but the potential exists.

It is questionable whether snake populations could support a widely used drug derived from venom. Most venom for biomedical research currently comes from wild-caught snakes. The volumes necessary to manufacture a popular drug would soon outstrip available resources, just as the commercialization of the anticancer drug taxol threatened the survival of the Pacific yew tree from which it was derived. Snakes can be raised in captivity, but not in the numbers necessary for pharmaceutical use. In this respect, the lack of emphasis on venom drugs serves conservation. The venom gathered during rattlesnake roundups is too contaminated for use in biomedical research or drug production.

WHY DO I HAVE SNAKES IN MY HOUSE?

A snake in the family room or on the doorstep is disconcerting for most people. Even urban apartment-dwellers sometimes receive visits from their neighbors' pet snakes or from wild species that thrive in urban areas.

The snake would prefer to be elsewhere, perhaps even more than you would prefer it to be elsewhere. Although three-quarters of the world's snake species are nonvenomous, and your unwelcome visitor probably belongs to one of them, it could be dangerous. Even an apartment-dweller cannot be sure that a neighbor does not keep venomous snakes (illegally, in most cities). In Florida, local venomous species such as the pygmy rattlesnake (*Sistrurus miliarius*), eastern diamondback rattlesnake (*Crotalus adamanteus*), and harlequin coralsnake (*Micrurus fulvius*) may appear in the backyard. Even the cottonmouth (*Agkistrodon piscivorus*) is seen around homes near swamps or lakes. Venomous visitors are also possible elsewhere in North America, although there are many more species of nonvenomous snakes than venomous ones.

For removal, call the local animal-control department. Such departments usually have a staff member or someone on call who can safely remove snakes. Few police departments or other emergency-service units have staff trained for snake removal. Nor are zoos, museums, or colleges equipped to assist in snake removals.

In suburban and rural areas, snake visitors are most common in late summer and early fall when snakes enter homes either in search of a safe winter haven or be-

When people build homes in natural areas, a snake on the doorstep is no less likely than a bear raiding their trash-cans. (rattlesnake; photo courtesy of the Tucson Herpetological Society)

cause the house is on its route to a hibernation site. Every year in the metropolitan Washington, D.C., area, we receive a rash of calls in mid-September to early October reporting small snakes in basements and ground-floor rooms. Almost without exception, they are young-of-the-year ringneck snakes (*Diadophis punctatus*). Ringnecks are totally harmless. They are probably searching for a hibernation site, but our knowledge of their basic biology is too incomplete to be certain. Other localities experience similar spring or autumn "migrations" of different species. In western Canada, red-sided gartersnakes (*Thamnophis sirtalis parietalis*) make both spring and autumn migrations through backyards as they move from hibernation sites to summer feeding grounds and back again.

Snake visitors in other seasons can signal the presence of an abundant population of mice and rats. Rodent-predator species linger and become temporary residents as long as the rodent population allows for easy hunting. Because most rodent predators are large, often more than 1 meter long, their presence does not long go unnoticed. Waterside homes and vacation cottages can also experience an abundance of fish- and frog-eating snakes in midsummer, when tadpoles aggregate at the water's edge.

The appearance of a snake visitor in your house or yard is usually a natural event that can alert you to a developing rodent problem, or it may just be a matter of two organisms using the same area simultaneously. It does not portend a snake infestation, or bad luck.

Ringneck snakes (*Diadophis punctatus*) are one of the several small snakes that accidentally enter houses in their autumn movements to hibernation sites. (G. Zug)

HOW DO I KEEP SNAKES OUT OF MY HOUSE AND YARD?

Many techniques and chemicals are proposed as snake repellents, but we know of none that is 100% effective. Most are only temporary deterrents, functioning at most for a few days to a few weeks.

First, we encourage a broad perspective. A "snake problem" usually arises because you have decided to move to the country or the woods to live closer to nature. Snakes and other so-called pests are just as much a part of nature as are woodthrushes and rabbits. Although it may be of little comfort, remember that you are invading the snake's space as much as it is invading yours.

Snakes visit because your house offers shelter or food or both. To discourage their arrival and short-term residence, you need to block access to shelter and food. Because rodents attract snakes, rodent control will reduce the likelihood of a snake invasion. Rodent control involves elimination of rodents' access to foodstuffs (particularly grains), removal of hiding places, and continual trapping.

Snakes gain access to houses through cracks and holes in foundations and ground-level walls, or through ill-fitting doors or ground-level windows. The first step is to remove all items stored against the house and to clear a 0.5- to 1-meter border between the shrubbery and the house, thus eliminating potential hiding places. Cracks and holes in the foundation and around doorjambs and window frames should then be sealed with concrete or a similar permanent material. Older homes often have loosely cemented fieldstone foundations. To make such houses snakeproof is prohibitively expensive, but careful do-it-yourself sealing of the foundation above and below ground level will discourage most snake visitors. Crawl spaces or larger spaces beneath the dwelling should not be used for storage.

In addition to maintaining a cleared border between house and shrubbery, dense ground cover of any sort should be eliminated or thinned; this landscaping, and regular trimming of grass, removes shelter. Woodpiles, compost heaps, and similar accumulations should be distant from the house. Firewood, lumber, and other materials should be stacked above open space and arranged to permit air circulation.

The most difficult snake problem occurs when one or more snakes takes up residence within the walls and ceilings of a house. Total fumigation is often promoted as the solution, but because there is no means of removing the dead animals, the owner must then tolerate the odor of decomposition. The best strategy may be to combine a capture-and-removal policy in the spring with sealing of the openings into the house and landscaping modifications.

Snake repellents such as bands of lime or mothballs, ammonia, pesticides, and petroleum products are sometimes recommended. These items are only temporarily effective as repellents, and they remain toxic. Thus they are likely to poison and kill plants and other animals, including pets.

Cats and goats are occasionally recommended for their ability to keep snakes away from homes. Cats are good predators, and they do catch and kill snakes, usually the small and inoffensive species. But cats kill many more songbirds and other wildlife than snakes. Goats, by contrast, are unlikely to attack and kill snakes. They repel snakes by removing vegetation and, in the case of billy goats, by their offensive odor—solutions likely to become more annoying than the original problem.

The introduction of mammalian predators as biological control agents has created major pest problems. Wherever stoats, ferrets, or mongoose have been introduced for purposes of rat or snake control, the introduced animal has become a pest and eliminated most of the native ground-living birds and mammals, while often having little effect on rats or snakes.

One repellent that appears effective is a specialized electric fence developed for the habu viper in the Amami Islands of southern Japan. The Okinawa habu (*Trimeresurus flavoviridis*) is abundant, and its bite, although now seldom lethal,

The brown treesnake causes numerous power failures when it short-circuits transmission lines by crawling across two electrical lines simultaneously. To avoid snake-caused power outages, the military airfield on Guam is encircled by a snake-proof fence. The bulge near the middle of the fence causes snakes to lose their balance and fall. (G. H. Rodda)

causes considerable pain and possible muscle damage. The islands have many small farms mixed with woods and steep hillsides, habitats that support numerous snakes. Biologists were enlisted to reduce the annual incidence of 250 to 300 habu bites. Detailed behavioral and ecological studies, combined with "snake-proofing" experiments, generated several effective methods for reducing human contact with habus. The first was rat control around dwellings. The second involved trapping habu using live rats or mice as bait. The third was the use of electrical fences, either of vinyl netting or cement blocks, at least 60 centimeters high. A stainless-steel band along the top of the fence is connected to a high-voltage pulse generator. Because of their expense, such fences are used only in selected areas. Snake fencing has also been developed for research on the effects of the brown treesnake (*Boiga irregularis*) on natural animal communities in Guam.

SHOULD I KEEP A SNAKE AS A PET?

We do not believe that any amphibian or reptile should be kept as a pet.

Why do we discourage keeping snakes when we are writing a book about them? Snakes are fascinating creatures, and we understand the desire to keep them in captivity. We also know many people who take good care of their snakes, but we have seen too much poor care. Too often, snakes and their kin are viewed as requiring minimum care rather than specialized care. The care provided is often inadequate, and the reptile or amphibian soon dies or experiences ongoing suffering.

Although some pet-trade amphibians and reptiles are now being bred in captivity, the majority still come from wild populations. During transport to pet stores, amphibians and reptiles are typically held and shipped in unsanitary and inhumane conditions. Many die, and most of the survivors arrive in ill health. Furthermore, collecting for the pet trade has drastically reduced many populations. Thus we view the pet trade mainly as an anti-conservation activity.

We realize that many people want snakes as pets, so even though we discourage the practice, we offer a few suggestions for keeping a pet snake healthy. Keeping a snake requires the same commitment and responsibility as keeping any other pet. You are responsible for its well-being and must be willing to expend the time and effort to ensure its proper care.

If this is your first snake, choose a common species that does not require specialized care or housing (ideally, a captive-bred individual) and about which adequate information exists. A well-cared-for pet snake will outlive a dog or cat. The life spans in captivity of ratsnakes and kingsnakes are commonly more than 15 years, and boas and pythons often live more than 20 years. Are you willing to commit yourself to maintain your snake for that long? Boa constrictors and pythons become large and potentially dangerous. Zoos rarely accept pet snakes, and finding a buyer—or even someone to accept a gift snake—can often be difficult.

Having decided what sort of snake you want, learn as much as possible about the species by reading the available literature before you buy it. You can also ask other snake keepers about the special needs of the species. With this knowledge, you can prepare for the arrival of the pet snake. This is also the time to check out the reliability of the pet stores or sellers you are considering.

Whether purchasing a snake or receiving one as a gift, check its health before accepting it. Caring for a sick snake will be frustrating and depressing, and will become expensive as well. The health of a snake can be difficult for a novice to determine, but a careful and thoughtful examination will offer some clues. Is the snake thin or emaciated? Are its scales wrinkled or dull? Are some scales raised rather than flat? Does it have scars or open sores? Are its eyes dull? Does it move oddly, yawn constantly, or wheeze? Does it have a discharge (liquid or semisolid) from its nose or mouth? Any one of these characteristics suggests that the snake is sick, and several definitely identify a sick snake.

Keeping your snake requires regular cleaning of its cage and regular feeding. Information on care is now readily available for many of the common snake species. We recommend reading several care manuals. If you consult several sources and use common sense, your snake should remain healthy and have a long life.

An exotic species might impress your friends, but its care requirements are probably poorly understood or unknown. Under these circumstances, even the best intentions do not serve the actual needs of the snake, and its health will begin to decline. *Maladaptation syndrome*—declining health because one or more life requirements are not met—reduces the snake's resistance to normal bacteria, which then become virulent, and the snake rapidly succumbs to septicemia, pneumonia, or similar diseases (see *Do Snakes Get Sick?*).

You might decide to catch a local snake for a pet. We discourage this practice, but we will also offer some hints for the well-being of the snake. As a pet, a larger-bodied species is best. Its food preference is likely to be easier to determine at your local library, and its food will be easier for you to catch or purchase.

If you have difficulty finding food for a local snake, you can return it to its home and allow it to fend for itself. Indeed, we recommend that local snakes be kept for only a few days or weeks and then returned to the wild, well in advance of the hibernation period (to ensure that the once-captive snake has enough time to build up adequate fat reserves for fuel during hibernation). When returning a local snake to nature, it must be released at the exact locality where it was captured. Doing so ensures the snake's familiarity with the locality, and thus its ability to find proper shelter and food. Otherwise the snake may search for its original home and die while searching. Also, moving a snake even a couple of kilometers might introduce it into a population with a differently adapted gene pool (heredity). If the released snake survives and reproduces, its genes might disrupt the local adaptation and decrease the probability of survival of subsequent generations.

Disease transmission is a serious hazard of keeping a local snake, no matter how briefly, and then returning it to its original home. A captive snake can pick up parasites and bacterial or viral diseases if housed with other snakes or in a cage previously used for other snakes. When released, it transmits the disease or parasites to the wild population. We know of no such transmission in snakes, but a respiratory disease transferred from pet Old World tortoises to native gopher tortoises now occurs in all North American tortoise species.

Venomous species as pets? *No!* A venomous snake as a pet is a contradiction in terms. Keeping a venomous snake in your home is like keeping a loaded and cocked handgun in your china cupboard. Sooner or later, someone will be seriously, perhaps even fatally, injured.

Keeping a venomous snake as a household pet is irresponsible. It endangers your life and the lives of family, friends, and neighbors. A bite of a pet venomous snake also threatens the lives of zoo personnel. Of the snake fanciers bitten by an exotic species in the metropolitan Washington area, none has had an adequate supply of antivenom serum on hand. The reptile departments of nearby zoos are expected to help, and of course they do. But their assistance reduces or eliminates their own

Ecologist must uniquely mark individuals to obtain demographic data. Snake biologist William S. Brown will clip the caudal scales to mark this timber rattler (*Crotalus horridus*). (C. Ernst)

stocks of antivenom serums, thus endangering the lives of their employees who work with venomous species.

HOW CAN I BECOME A SNAKE SPECIALIST?

A career working with snakes is possible, but competition is intense. A well-educated candidate with a variety of skills is most likely to get the job.

Until recently, most snake specialists worked as professors or researchers in colleges and universities. Such opportunities remain, but today herpetologists need expertise in a broader field, such as physiology, behavior, ecology, or molecular genetics for which they can use snakes as research models. There now exists a need for veterinarians with expertise in exotic animals, including reptiles. Other opportunities exist in environmental sciences, both in government and in private industry. All of these job opportunities require a college degree, and most also require an advanced degree. Few if any such jobs are exclusively devoted to snakes. Even the animal keepers in a zoo reptile house have many other responsibilities besides feeding snakes.

If you can combine your interest in snakes with another skill, you might be able to develop a career as a nature photographer, illustrator, journalist, or nature-tour leader. Once again, your chance of success depends on your education and ability to adapt your skills to a variety of tasks.

WHAT REMAINS TO BE LEARNED ABOUT SNAKES?

From cell and tissue function to behavior and ecology, we are only beginning to discover the complexities of snake biology. Biological study of snakes began about 300 years ago in the early eighteenth century, when naturalists first began to record and test observations on plants and animals. Much was already known about snakes, extending well back in time to the scholarly discoveries of ancient China, Egypt, Greece, and other civilizations. The problem with the knowledge that survived in books and manuscripts was that it combined actual observations with fallacious interpretation and downright flights of fancy.

The concept of repeated and repeatable observations was not introduced into research until the scientific revolution of the seventeenth century, and did not become the standard of acceptance until the nineteenth century. In modern science—be it snake biology, human medicine, or astronomy—an observation is not simply believed; it is tentatively accepted until a new observation, test, or experiment suggests that it is incorrect. Repeated observation and experimentation, as well as questioning and reexamining observations and interpretations, are essential features of science. Thus, for example, creationism (or its current disguise: intelligent-design theory) does not qualify as science because it is not based on this test-and-modify principle.

The concepts that we accept about snakes and their biology require repeated testing in the light of new discoveries and techniques. Throughout this book, we have noted matters about which we are uncertain because our knowledge is incomplete. Among these grey areas are questions about how the snakes evolved from lizards, the evolution of modern snakes, and the relationships of modern snakes. The fossil record is incomplete and of little help. Biochemical and molecular data are beginning to be integrated with data on anatomy and physiology, and are providing answers about relationships.

Many aspects of snake physiology remain poorly understood, including such critical processes as gas exchange through a reduced lung surface and blood pressure at various sites in the circulatory system. Processes involved in snakes' responses to temperature, including resistance to freezing during hibernation and the roles of temperature and hydration on sex determination (particularly in live-bearing snakes) remain unexplored. Finally, there is the intriguing question of why some species of snakes are immune to their own venom or that of other snakes and others are not.

Our knowledge of snake physiology lags far behind our knowledge of human physiology. Humans, as diverse as we are, represent only one species—snakes comprise more than 2,800. Even so, the number of biomedical researchers (studying human physiology) far exceeds the number of biologists studying all other animal species. Similarly, all other aspects of snake biology remain only superficially known, leaving many challenges for future investigators.

APPENDIX 1

CLASSIFICATION OF SNAKES

Hierarchical Classification of Snake Families[a]

Reptilia
  Sauria
    Lepidosauria
      Squamata (lizards)
        Autarchoglossa
          Diploglossa
            Anguimorpha
              Thecoglossa (monitors & snakes)
                Varanidae (monitors)
                Pythonomorpha (ancient & modern snakes)
                  SERPENTES (snakes)
                    Scolecophidia
                      Leptotyphlopidae
                      unnamed clade
                        Anomalepididae
                        Typhlopidae
                  Alethinophidia
                    Anomochilidae
                      Uropeltidae
                        Cylindrophiidae
                      Aniliidae
                      Xenopeltidae
                      unnamed clade
                      Loxocemidae
                      henophidians
                      Boidae
                        Boinae
                        Erycinae
                    Pythonidae
                    unnamed clade
                    Bolyeriidae
                    unnamed clade
                      Tropidophiidae
                        Tropidophiinae
                        Ungaliophiinae
                        Xenophidioninae
                    Caenophidia

(continued)

Acrochordidae
colubroids
   Viperidae
      Azemiopinae
      Crotalinae
      Viperinae
   unnamed clade
      Atractaspididae
         Aparallactinae
         Atractaspidinae
      Colubridae
         Colubrinae
         Homalopsinae
         Lamprophiinae
         Natricinae
         Pareatinae
         Xenodermatinae
         Xenodontinae
      Elapidae
         Elapinae
         Hydrophiinae

---

[a] The classification derives largely from the one presented in Zug, Vitt, and Caldwell 2001. Each indentation represents an evolutionary divergent event; however, we have not attempted to include all events in the evolution of snakes from the first reptile ancestor. The occurrence of three or more taxa at the same level of indentation indicates an uncertainty of evolutionary history.

Representative Genera of the Families and Subfamilies of Snakes[b]

Family and subfamily	Number of genera	Number of species	Select genera
SCOLECOPHIDIANS			
Anomalepididae (early blindsnakes)	4	15	*Anomalepis, Helminthophis, Liotyphlops, Typhlophis*
Leptotyphlopidae (threadsnakes or wormsnakes)	2	90+	*Leptotyphlops, Rhinoleptus*
Typhlopidae (blindsnakes)	6	210+	*Ramphotyphlops, Rhinotyphlops, Typhlops*
BASAL ALETHINOPHIDIANS			
Aniliidae (false coralsnake)	1	1	*Anilius*
Anomochilidae (false blindsnakes)	1	2	*Anomochilus*
Boidae			
Boinae (boas)	6	25+	*Boa, Candoia, Corallus, Epicrates, Eunectes, Sanzina*
Erycinae (sandboas)	2	14	*Charina, Eryx*
Bolyeriidae (Mascarene boas)	2	2	*Bolyeria, Casarea*
Cylindrophiidae (pipesnakes)	1	8	*Cylindrophis*
Loxocemidae (Mesoamerican python)	1	1	*Loxocemus*
Pythonidae (pythons)	4	25+	*Aspidites, Liasis, Morelia, Python*
Tropidophiidae			
Tropidophiinae (dwarfboas)	2	18	*Trachyboa, Tropidophis*
Ungaliophiinae	2	3	*Exiliboa, Ungaliopia*
Xenophidioninae (Asian dwarfboas)	1	2	*Xenophidion*
Uropeltidae (shieldtails or shield-tailed snakes)	9	45+	*Plectrurus, Rhinophis, Uropeltis*
Xenopeltidae (sunbeam snake)	1	2	*Xenopeltis*
CAENOPHIDIANS			
Acrochordidae (filesnakes or wartsnakes)	1	3	*Acrochordus*
Atractaspididae			
Aparallactinae (glossysnakes)	12	40+	*Amblyodipsas, Aparallactus, Macrelaps, Polemon, Xenocalamus*
Atractaspidinae (molevipers or stilettovipers)	1	17	*Atractaspis*
Colubridae			
Colubrinae (colubrine snakes)	100+	650+	*Ahaetulla, Arizona, Bogertophis, Boiga, Calamaria, Cemophora, Chionactis, Chironus, Chrysopelea, Coluber, Coronella, Dasypeltis, Dendrelaphis, Dispholidus, Drymarchon, Elachistodon, Elaphe, Ficimia, Gonyosoma, Gyalopion, Lampropeltis, Lycodon, Malpolon, Masticophis, Mastigodryas, Opheodrys, Pituophis, Prosymna, Pseustes, Ptyas, Rhinocheilus, Salvadora, Senticolis, Sonora, Stilosoma, Tantilla, Thelotornis, Trimorphodon*
Homalopsinae (rear-fanged watersnakes)	11	35+	*Cerberus, Enhydris, Erpeton, Fordonia, Homalopsis, Myron*
Lamprophiinae (lamprophiine snakes)	44	205+	*Boaedon, Dendrolycus, Duberria, Grayia, Lamprophis, Lycophidion, Mehelya, Mimophis, Polemon, Pseudaspis, Scaphiophis*

(continued)

Family and subfamily	Number of genera	Number of species	Select genera
Natricinae (watersnakes)	38	200+	*Amphiesma, Aspidura, Atretium, Clonophis, Natrix, Nerodia, Rhabdophis, Seminatrix, Storeria, Thamnophis, Virginia, Xenochrophis*
Pareatinae (slugeaters)	3	18+	*Aplopeltura, Internatus, Pareas*
Xenodermatinae (Asian forest snakes)	6	15	*Achalinus, Fimbrios, Oxyrhabdium, Xenodermus*
Xenodontinae (xenodontine snakes)	90+	540+	*Adelphicos, Alsophis, Atractus, Carphophis, Clelia, Coniophanes, Diadophis, Dipsas, Erythrolamprus, Farancia, Geophis, Helicops, Heterodon, Hypsiglena, Imantodes, Leptodeira, Liophis, Philodryas, Pliocercus, Rhadinaea, Sibon, Tachymenis, Uromacer, Urotheca, Xenodon*
Elapidae			
Elapinae (cobras, coralsnakes, and allies)	17	130+	*Aspidelaps, Boulengerina, Bungarus, Calliophis, Dendroaspis, Maticora, Micruroides, Micrurus, Naja, Ophiophagus, Pseudohaje*
Hydrophiinae (tigersnakes, seasnakes)	43	165+	*Acanthophis, Astrotia, Austrelaps, Demansia, Enhydrina, Hemiaspis, Hydrophis, Lapemis, Laticauda, Notechis, Ogmodon, Oxyuranus, Pelamis, Pseudonaja, Simoselaps, Suta*
Viperidae			
Azemiopinae (Fea's viper)	1	1	*Azemiops*
Crotalinae (pitvipers)	18	155+	*Agkistrodon, Bothriechis, Bothrops, Calloselasma, Crotalus, Deinagkistrodon, Hypnale, Lachesis, Ophryacus, Ovophis, Porthidium, Sistrurus, Trimeresurus, Tropidolaemus*
Viperinae (true vipers)	13	65+	*Atheris, Bitis, Causus, Cerastes, Echis, Eristocophis, Pseudocerastes, Vipera*

[b] We have attempted to follow the most current nomenclature (scientific and vernacular); with a few minor exceptions, names follow the usage of Crother 2000, David and Ineich 1999, McDiarmid et al. 1999, and Zug et al. 2001.

APPENDIX 2

SLOWER THAN THEY SEEM: LOCOMOTION SPEEDS OF SELECT SNAKES AND OTHER ANIMALS

Maximum speed (kilometers/hour)	Mode of locomotion	Taxon
Running or crawling		
80–98	Burst	Cheetah
56–64	Fast cruise	Jackrabbit
42–48	Gallop	Race horse
32–38	Sprint	Human
16.00[a]	Serpentine	Kirtland's birdsnake (*Thelotornis kirtlandii*)
11.26	Serpentine	Black mamba (*Dendroaspis polylepis*)
5.96	Serpentine	Eastern racer (*Coluber constrictor*)
5.80	Serpentine	Coachwhip (*Masticophis flagellum*)
3.28	Sidewinding	Sidewinder rattlesnake (*Crotalus cerastes*)
1.90	Serpentine	Pinesnake (*Pituophis melanoleucus*)
1.61	Serpentine	African rock python (*Python sebae*)
1.21	Serpentine	Glossysnake (*Arizona elegans*)
0.65	Serpentine	Mojave rattlesnake (*Crotalus scutulatus*)
0.60	Serpentine	Eastern ratsnake (*Elaphe obsoleta*)
0.36	Serpentine	Rosy boa (*Lichanura trivirgata*)
0.10	Rectilinear	Indian python (*Python molurus*)
Swimming		
26–32	Burst	Sperm whale
20–45	Burst	Tuna
4.09	Undulatory	Southern watersnake (*Nerodia fasciata*)
2.59	Undulatory	Cornsnake (*Elaphe guttata*)
Gliding or flying		
38.74	Controlled fall	Golden flyingsnake (*Chrysopelea ornata*)
14–18[b]	Active flying	Little brown bat

[a]This speed is not based on a field experiment and is probably incorrect.
[b]This value represents a normal, not a maximum, speed.

APPENDIX 3

HERPETOLOGICAL ORGANIZATIONS

The twentieth century began without a single organization devoted to the study of amphibians and reptiles. Now in the twenty-first century, hundreds exist. Societies or groups range from small groups of herp fanciers in a single town to international organizations that have a thousand or more members. This list includes only a sampling of those in the United States.

Because societies are dependent on volunteers to serve as officers, mailing addresses change frequently. The following addresses were correct when we wrote this book but since may have changed. We encourage the reader to surf the web because many herpetological organization maintain web pages. We purposefully have not listed web addresses because these have even a shorter "shelf-life" than do the business offices of the larger organizations.

American Society of Ichthyologists and Herpetologists (ASIH)
ASIH Business Office
810 East 10th Street
PO Box 1897
Lawrence, KS 66044-2001

Herpetologists' League (HL)
c/o M. F. Given
Department of Biology
Neumann College
One Neumann Drive
Aston, PA 19014-1298

Society for the Study of Amphibians and Reptiles (SSAR)
c/o D. Schmitt
P.O. Box 253
Marceline, MO 64658-0253

Select Regional Societies within the United States
Chicago Herpetological Society
Membership Secretary
2060 N Clark Street
Chicago, IL 60614

Gainesville Herpetological Society
PO Box 140353
Gainesville, FL 32614–0353

Kansas Herpetological Society
5438 SW 12th Terrace, Apt. 4
Topeka, KS 66604

Maryland Herpetological Society
Department of Herpetology
Natural History Society of Maryland
2643 N Charles Street
Baltimore, MD 21218

Minnesota Herpetological Society
Bell Museum of Natural History
10 Church Street SE
Minneapolis, MN 55455-0104

New England Herpetological Society
PO Box 1082
Boston, MA 02103

North Carolina Herpetological Society
N.C. State Museum of Natural Sciences
11 West Jones Street
Raleigh, NC 27601-1029

Northern Ohio Association of Herpetologists (NOAH)
Department of Biology
Case Western Reserve University
Cleveland, OH 44106-7080

San Diego Herpetological Society
PO Box 4036
San Diego, CA 92164-4036

Southwestern Herpetological Society
PO Box 7469
Van Nuys, CA 91409-7469

Tucson Herpetological Society
PO Box 709
Tucson, AZ 85702-0709

Virginia Herpetological Society
c/o P. Sattler
Department of Biology
Liberty University
1971 University Boulevard
Lynchburg, VA 24502

GLOSSARY

Synonyms and alternative spellings are in parentheses; different parts of speech are in square brackets.

adaptation Any structure, physiological process, or behavior trait that gives an organism a better chance to survive and reproduce in its environment.

aggregation A gathering of organisms, usually for a social or physiological purpose.

aglyphous Possessing teeth without grooves or a closed canal.

allantois An embryonic membrane that allows an embryo to breathe and to dispose of waste products.

amnion An embryonic membrane that surrounds the embryo and encloses it in a fluid-filled sac.

amniotic A vertebrate embryo developing in a fluid-filled sac surrounded by an amniotic membrane.

anal plate A single or paired scale situated in front of the vent.

anterior Toward the front of an animal.

antivenom (antivenin) A serum, prepared from the blood of horses or sheep, used to neutralize the effects of envenomation by venomous snakes.

apnea A pause in breathing.

aposematic [noun, aposematism] Characterized by bright or contrasting coloration, serving to warn off potential predators.

brooding Physical contact between a mother and her eggs for part or all of the eggs' incubation period.

brumation Torpor without physiological adjustment. *Compare* hibernation.

button The terminal segment on the tail rattle of a rattlesnake.

caenophidian An "advanced" snake (i.e., a member of suborder Caenophidia), such as a colubrid, elapid, or viperid; *see* Appendix 1.

caudal Toward the tail or rear of the body.

chromatophore (pigment cell) Any of several types of cells that carry pigment, including iridophores, melanophores, and xanthophores.

cloaca A body chamber that receives the products of the digestive, urinary, and reproductive systems immediately prior to their exit from the body.

clutch The eggs laid by a female at one nesting.

colubrid A snake of the family Colubridae.

combat dance A pushing–shoving contest between a pair of male snakes to determine sexual dominance.

competition A demand by two or more individuals for the same resource (food, water, space, shelter, mates).

cornified *See* keratinous.

cranial [noun, cranium] Of the head, usually excluding the lower jaw.

crepuscular Active at dusk, at dawn, or both.

Cretaceous The last period of the Mesozoic Era, or Age of the Reptiles.

dermal [noun, dermis] Pertaining to the sublayer of the skin containing the blood vessels and nerves.

desiccation (dehydration) Loss of water from body tissues and cells, usually across the skin and lung surfaces in snakes.

diastema An empty (toothless) space in a row of teeth on the same bone.

distal Toward the tip of a limb or other protruding structure.

diurnal Active during daylight.

dorsal [noun, dorsum] Of or on the back.

Duvernoy's gland (parotid gland) A portion of the superior labial salivary gland in nonvenomous colubrid snakes that corresponds to the venom gland in venomous snakes.

ecdysis (molting, shedding) Periodic shedding of the entire skin.

ectothermy [adjective, ectothermic] Dependence on an external heat source to elevate body temperature.

egg tooth A true tooth on the premaxillary of prehatchling snakes, used to slit the eggshell during hatching.

endemic Native to a particular area and found nowhere else.

endothermy [adjective, endothermic] Production of body heat internally through cellular metabolism.

envenomate To inject venom into the puncture wound resulting from bite or sting.

enzyme A protein that brings about or intensifies a specific chemical reaction.

epidermal [noun, epidermis] Of the outer layer of the skin.

estivation Inactivity and torpor during a hot and/or dry period; the opposite of hibernation.

evolution Change in a population over time; change in a population's genetic structure, expressed by differences in behavior, anatomy, physiology, and other traits.

excretory Of the urinary system, including kidneys and associated ducts and their products.

facial pit (loreal pit) *See* pit organ.

facultative Capable of but not obligated to a specific physiological process, such as hibernation, or a specific mode of life, such as aquatic.

fang A tooth with a groove or closed canal that channels venom into a bite.

fossorial Burrowing; usually, a completely subterranean life style.

gastrosteges (ventral or belly scales) Large transverse scales on the underside of a snake's body.

generalist An animal whose diet encompasses a variety of foods.

glottis A structure on the floor of the rear of the mouth cavity bearing the opening into the trachea.

hemipenes [singular, hemipenis] The paired male copulatory organs of lizards and snakes.

hemotoxic (hematotoxic, haemotoxic, hemolytic) Destructive to blood and circulatory tissues; loosely used to refer to destruction of any tissue other than nervous tissue.

henophidian An "early" snake (i.e., a member of suborder Henophidia), such as a boa, python, or one of their distant relatives.

herpetology The study of the biology and evolution of amphibians and reptiles.

hertz A measure of sound frequency in cycles per second; low numbers represent the low or bass sounds; high numbers, high treble or high-pitched sounds.

hibernaculum A den or other site shared by a group of hibernating animals.

hibernation Inactivity or torpor associated with a cold season, typically involving physiological adjustments.

homoiothermic [noun, homoiothermy] Maintaining a constant body temperature.

hybridization Reproduction between male and female of different species.

integument The external covering or skin of an animal. In vertebrates, the integument contains the epidermis and dermis.

Jacobson's organ (vomeronasal organ) An olfactory organ located in the roof of the mouth near the internal nostrils.

juxtaposed Lying side by side without overlapping.

keeled scale (carinate scale) A scale bearing a longitudinal ridge or ridges.

keratinous (cornified, horny) [noun, keratin] Possessing a dense protein that forms the outer layer of the skin in many vertebrates, including reptilian scales, bird feathers, mammal hair, and the claws and nails of reptiles, birds, and mammals.

kinetic [noun, kinesis] Mobile or flexible; commonly used to describe bending or hinged joints in the skeleton.

labial pits Sensory pits located between the upper and lower lip scales of some boas and pythons.

lateral Toward or on the sides of the body.

LD$_{50}$ A measure of the toxicity of a substance such as venom; the lethal dose, or the amount of the substance that is required to kill 50% of the animals in the test sample within 24 hours (in the case of venom, reported as milligrams of venom per kilogram of mice).

letisimulation Death-feigning, or "playing dead," as a defensive behavior.

litter The offspring of a multiple birth.

lumen An opening or chamber in a tissue or organ.

maladaptation syndrome Poor and declining health in captive animals, resulting from improper care and nutrition.

mandible The lower jaw.

maxilla (maxillary bone) A major bone of the upper jaw, which in venomous snakes bears the fang.

melanophore A pigment cell carrying the dark-brown pigment melanin.

milking Extraction of venom.

nasal Associated with the nose and the adjacent area of the head.

neurotoxic (neurolytic) Destructive to nerve-impulse transmission within and between nerves and muscles.

nictitating membrane A transparent eyelid-like structure, which lies beneath the primary or external eyelids.

nocturnal Active at night.

Oberhautchen The outermost layer of the epidermis.

obligatory Obligated to a specific physiological process, such as hibernation, or a specific mode of life, such as aquatic, in order to survive.

oophagous Egg-eating.

ophiophagous Snake-eating.

opisthoglyphous Possessing enlarged teeth, either grooved or ungrooved, on the rear half of the maxillary bone.

oviparous [noun, oviparity] Producing fertilized eggs that develop and hatch outside of the female's body.

ovoviviparous [noun, ovoviviparity] Producing fertilized eggs that are retained in the female's oviduct (uterus) but develop without a placental attachment to the female.

parthenogenesis [adjective, parthenogenetic] Clonal reproduction; production of self-activated (self-fertilized) eggs that produce female offspring with the same genetic constitution as that of the mother.

pelagic Living in the open ocean.

photoreceptor A cell or tissue that is sensitive to light.

piscivorous Fish-eating.

pit organ A sensory organ, located between the nostril and eye of one group of viperid snakes, that is capable of sensing infrared radiation.

Pleistocene The earliest epoch of the Quaternary Period.

poikilothermic [noun, poikilothermy] Possessing an inconstant body temperature that varies occasionally or continually with the environmental temperature.

poison Any substance that irritates or kills.

posterior Toward or on the rear of an animal.

proteroglyphous Possessing fixed fangs in the front of the mouth.

proximal Toward the base or attachment of a limb or other protruding structure.

Quaternary The last (current) period of the Cenozoic Era.

rattle A terminal scaly appendage on the tail of rattlesnakes.

rectilinear locomotion Slow movement by shifting the body forward within the skin, also known as caterpillar locomotion or "walking on the ribs."

scalation (scutellation) The number and pattern of a snake's scales.

shaman A tribal priest who heals ills by removing evil spirits.

sheath A membrane enclosing or covering a structure. The fang sheath surrounding the base of a fang helps funnel venom into the fang's venom canal.

smooth scale A scale lacking longitudinal ridges or keels.

solenoglyphous Possessing hinged fangs in the front of the mouth.

specialist An animal whose diet is restricted to specific foods.

spectacle The transparent scale covering the eye of snakes and some lizards.

sternum The breast bone.

substrate The surface on which an animal lives and moves.

taxon [plural, taxa] Any group of organisms that is considered distinct from other groups, such as reptiles, snakes, pythons, or the eastern black ratsnake.

temporal The area or bone on the side of the head behind the eye.

Tertiary The first period of the Cenozoic Era.

thanatosis *See* letisimulation.

thermal conformist An ectotherm that allows its body temperature to vary with the temperature of its surroundings.

thermal nonconformist An ectotherm that attempts to attain and maintain a constant body temperature.

vascular Pertaining to the blood system; possessing numerous blood vessels.

venom A poisonous substance injected into a wound by bite or sting.

vent The opening of the cloaca; distinct from the anus, which is the opening of the digestive system into the cloaca.

ventral [noun, venter] Of or on the underside of an animal.

vertebrate An animal with a backbone (e.g., fish, amphibians, reptiles, birds, mammals).

viviparous [noun, viviparity] Producing fertilized eggs that are retained in the female's oviduct (uterus) and develop with a placental attachment to the female.

vomeronasal organ *See* Jacobson's organ.

yolk sac The membrane surrounding the yolk supply of an embryo, which late in development is retracted and becomes part of the wall of the digestive tract.

GENERAL BIBLIOGRAPHY

Bauchot, R., ed. 1994. *Snakes: A Natural History*. New York: Sterling.

Branch, W. 1991. *Everyone's Guide to Snakes of Southern Africa*. Johannesburg: Central News Agency.

Campbell, J. A., and W. W. Lamar. 1989. *The Venomous Reptiles of Latin America*. Ithaca, New York: Comstock Publishing Associates, Cornell University Press.

Cogger, H. G., and R. G. Zweifel, eds. 1998. *Encyclopedia of Reptiles & Amphibians*. 2nd Ed. San Diego: Academic Press.

Conant, R., and J. T. Collins. 1991. *A Field Guide to Reptiles and Amphibians: Eastern and Central North America*. Boston: Houghton Mifflin.

Engelmann, W., and F. J. Obst. 1981. *Snakes: Biology, Behavior, and Relationships to Man*. New York: Exeter Books.

Ernst, C. H., and E. M. Ernst. 2003. *Snakes of the United States and Canada*. Washington: Smithsonian Institution Press.

Fitch, H. S. 1999. *A Kansas Snake Community: Composition and Changes over 50 Years*. Malabar, Florida: Krieger.

Greene, H. W. 1997. *Snakes: The Evolution of Mystery in Nature*. Berkeley: University of California Press.

Greer, A. E. 1998. *The Biology and Evolution of Australian Snakes*. Chipping Norton, New South Wales: Surrey Beatty & Sons.

Halliday, T., and K. Adler, eds. 2002. *The New Encyclopedia of Reptiles and Amphibians*. Oxford: Oxford University Press.

Heatwole, H. 1999. *Sea Snakes*. Malabar, Florida: Krieger.

Ineich, I., and P. Laboute. 2002. *Les Serpents marins de Nouvelle-Calédonie. Sea snakes of New Caledonia*. Paris: IRD Éditions.

Klauber, L. M. 1972. *Rattlesnakes: Their Habits, Life Histories, and Influence on Mankind*, 2nd Ed. Berkeley: University of California Press.

Minton, S. A., Jr., and M. R. Minton. 1973. *Giant Reptiles*. New York: Charles Scribner's Sons.

———. 1980. *Venomous Reptiles*, rev. ed. New York: Charles Scribner's Sons.

Parker, H. W., and A. G. C. Grandison. 1977. *Snakes: A Natural History*, 2nd Ed. Ithaca, New York: Cornell University Press.

Roze, J. A. 1996. *Coral Snakes of the Americas: Biology, Identification, and Venoms*. Malabar, Florida: Krieger.

Seigel, R. A., and J. T. Collins, eds. 1993. *Snakes: Ecology and Behavior*. New York: McGraw-Hill.

Seigel, R. A., J. T. Collins, and S. S. Novak, eds. 1987. *Snakes: Ecology and Evolutionary Biology*. New York: Macmillan.

Shine, R. 1991. *Australian Snakes: A Natural History*. Ithaca, New York: Cornell University Press.

Zug, G. R., L. J. Vitt, and J. P. Caldwell. 2001. *Herpetology: An Introductory Biology of Amphibians and Reptiles*. 2nd Ed. San Diego: Academic Press.

SUBJECT BIBLIOGRAPHY

SNAKE FACTS

What Are Snakes?

Bauchot, R., ed. 1994. *Snakes: A Natural History*. New York: Sterling.

Edwards, J. L. 1985. Terrestrial locomotion without appendages. In *Functional Vertebrate Morphology*, edited by M. Hildebrand et al., 159–172. Cambridge, Massachusetts: Harvard University Press.

Ernst, C. H., and E. M. Ernst. 2003. *Snakes of the United States and Canada*. Washington: Smithsonian Institution Press.

Gans, C. 1974. *Biomechanics: An Approach to Vertebrate Biology*. Philadelphia: Lippincott.

———. 1975. Tetrapod limblessness: Evolution and functional corollaries. *American Zoologist* 15:455–467.

Gauthier, J., A. G. Kluge, and T. Rowe. 1988. Amniote phylogeny and the importance of fossils. *Cladistics* 4:105–209.

Halliday, T., and K. Adler, eds. 2002. *The New Encyclopedia of Reptiles and Amphibians*. Oxford: Oxford University Press.

Parker, H. W., and A. G. C. Grandison. 1977. *Snakes: A Natural History*, 2nd Ed. Ithaca, New York: Cornell University Press.

Zug, G. R., L. J. Vitt, and J. P. Caldwell. 2001. *Herpetology: An Introductory Biology of Amphibians and Reptiles*. 2nd Ed. San Diego: Academic Press.

How Are Snakes Built?

Cundall, D. 1987. Functional morphology. In *Snakes: Ecology and Evolutionary Biology*, edited by R. A. Seigel, J. T. Collins, and S. S. Novak, 105–140. New York: Macmillan.

Gomes, N., G. Puorto, M. A. Buononato, and M. de F. M. Ribeiro. 1989. Atlas anatômico de *Boa constrictor* Linnaeus, 1758. *Instituto Butantan Monograph* 2:1–59.

Hoffstetter, R., and J. P. Gasc. 1969. Vertebrae and ribs of modern reptiles. In *Biology of the Reptilia*, Vol. 1, edited by C. Gans, A. d'A. Bellairs, and T. S. Parsons, 201–311. London: Academic Press.

Lillywhite, H. B. 1993. Subcutaneous compliance and gravitational adaptations in snakes. *Journal of Experimental Zoology* 267:557–562.

Oldham, J. C., H. M. Smith, and S. A. Miller. 1970. A *Laboratory Perspectus of Snake Anatomy*. Champaign, Illinois: Stipes.

Parker, H. W., and A. G. C. Grandison. 1977. *Snakes: A Natural History*, 2nd Ed. Ithaca, New York: Cornell University Press.

Underwood, G. 1967. A *Contribution to the Classification of Snakes*. London: British Museum (Natural History).

Zippel, K. C., H. B. Lillywhite, and C. R. J. Mladinich. 2001. New vascular system in reptiles: anatomy and postural hemodynamics of the vertebral venous plexus in snakes. *Journal of Morphology* 250:173–184.

How Do Snakes Breathe?

Baeyens, D. A., M. W. Patterson, and C. T. McAllister. 1980. A comparative physiological study of diving in three species of *Nerodia* and *Elaphe obsoleta*. *Journal of Herpetology* 14:65–70.

Clark, B. D., C. Gans, and H. I. Rosenberg. 1978. Air flow in snake ventilation. *Respiration Physiology* 32:207–212.

Heatwole, H., and R. Seymour. 1976. Respiration of marine snakes. In *Respiration of Amphibious Vertebrates*, edited by G. M. Hughes, 375–389. London: Academic Press.

Seymour, R. S. 1982. Physiological adaptations to aquatic life. In *Biology of the Reptilia*, Vol. 13, edited by C. Gans and F. H. Pough, 1–51. London: Academic Press.

Wallach, V. 1998. The lungs of snakes. In *Biology of the Reptilia*, Vol. 13, edited by C. Gans and A. Gaunt, 93–295. Ithaca, New York: Society for the Study of Amphibians and Reptiles.

How Do Snakes Crawl and Swim?

Cundall, D. 1987. Functional morphology. In *Snakes: Ecology and Evolutionary Biology*, edited by R. A. Seigel, J. T. Collins, and S. S. Novak, 105–140. New York: Macmillan.

Edwards, J. L. 1985. Terrestrial locomotion without appendages. In *Functional Vertebrate Morphology*, edited by M. Hildebrand et al., 159–172. Cambridge, Massachusetts: Harvard University Press.

Gans, C. 1974. *Biomechanics: An Approach to Vertebrate Biology*. Philadelphia: Lippincott.

———. 1986. Locomotion of limbless vertebrates: Patterns and evolution. *Herpetologica* 42:33–46.

Lillywhite, H. B., and R. W. Henderson. 1993. Behavioral and functional ecology of arboreal snakes. In *Snakes: Ecology and Behavior*, edited by R. A. Seigel and J. T. Collins, 1–48. New York: McGraw-Hill.

How Do Snakes See?

Barrett, R. 1970. The pit organs of snakes. In *Biology of the Reptilia*, Vol. 2, edited by C. Gans and T. S. Parsons, 277–300. London: Academic Press.

Ford, N. B., and G. M. Burghardt. 1993. Perceptual mechanisms and the behavioral ecology of snakes. In *Snakes: Ecology and Evolutionary Biology*, edited by R. A. Seigel and J. T. Collins, 117–164. New York: McGraw-Hill.

Molenaar, G. J. 1992. Anatomy and physiology of infrared sensitivity of snakes. In *Biology of the Reptilia*, Vol. 17, edited by C. Gans and P. S. Ulinski, 367–435. Chicago: University of Chicago Press.

Newman, E. A., and P. H. Hartline. 1982. The infrared "vision" of snakes. *Scientific American* 246(3):116–125.

Underwood, G. 1970. The eye. In *Biology of the Reptilia*, Vol. 2, edited by C. Gans and T. S. Parsons, 1–97. London: Academic Press.

Do Snakes Hear?

Harline, P. H. 1971. Physiological basis for detection of sound and vibration in snakes. *Journal of Experimental Biology* 54:349–371.

Wever, E. G. 1978. *The Reptile Ear: Its Structure and Function*. Princeton, New Jersey: Princeton University Press.

Young, B. A. 1997. A review of sound production and hearing in snakes, with a discussion of intraspecific acoustic communication in snakes. *Journal of the Pennsylvania Academy of Science* 71:39–46.

Can Snakes Smell?

Ford, N. B., and G. M. Burghardt. 1993. Perceptual mechanisms and the behavioral ecology of snakes. In *Snakes: Ecology and Evolutionary Biology*, edited by R. A. Seigel and J. T. Collins, 117–164. New York: McGraw-Hill.

Halpern, M. 1992. Nasal chemical senses in reptiles: Structure and function. In *Biology of the Reptilia*, Vol. 18, edited by C. Gans and D. Crews, 423–523. Chicago: University of Chicago Press.

Parsons, T. S. 1970. The nose and Jacobson's organ. In *Biology of the Reptilia*, Vol. 2, edited by C. Gans and T. S. Parsons, 99–191. London: Academic Press.

Schwenk, K. 1994. Why snakes have forked tongues. *Science* 263:1573–1577.

———. 1995. Of tongues and noses: Chemoreception in lizards and snakes. *Trends in Ecology and Evolution* 10:7–12.

Do Snakes Taste Their Food?

Schwenk, K. 1994. Why snakes have forked tongues. *Science* 263:1573–1577.

Young, B. A. 1997. On the absence of tastebuds in monitor lizards (*Varanus*) and snakes. *Journal of Herpetology* 31:130–137.

How Do Snakes Find and Subdue Their Prey?

Arnold, S. J. 1993. Foraging theory and prey-size–predator-size relations in snakes. In *Snakes: Ecology and Behavior*, edited by R. A. Seigel and J. T. Collins, 57–115. New York: McGraw-Hill.

Cundall, D. 1987. Functional morphology. In *Snakes: Ecology and Evolutionary Biology*, edited by R. A. Seigel, J. T. Collins, and S. S. Novak, 105–140. New York: Macmillan.

Greene, H. W. 1997. *Snakes: The Evolution of Mystery in Nature*. Berkeley: University of California Press.

Hardy, D. L. 1994. A re-evaluation of suffocation as the cause of death during constriction by snakes. *Herpetological Review* 25:45–47.

Moon, B. R. 2000. The mechanics and muscular control of constriction in gopher snakes (*Pituophis melanoleucus*) and a king snake (*Lampropeltis getula*). *Journal of the Zoological Society, London* 252:83–98.

Mushinsky, H. R. 1987. Foraging ecology. In *Snakes: Ecology and Evolutionary Biology*, edited by R. A. Seigel, J. T. Collins, and S. S. Novak, 302–334. New York: Macmillan.

What and How Do Snakes Eat?

Cundall, D. 1987. Functional morphology. In *Snakes: Ecology and Evolutionary Biology*, edited by R. A. Seigel, J. T. Collins, and S. S. Novak, 105–140. New York: Macmillan.

Cundall, D., and C. Gans. 1979. Feeding in water snakes: an electromyographic study. *Journal of Experimental Zoology* 209:189–208.

Cundall, D., and H. W. Greene. 2000. Feeding in snakes. In *Feeding: Form, Function, and Evolution in Tetrapod Vertebrates*, edited by K. Schwenk, 293–333. San Diego: Academic Press.

Ernst, C. H., and E. M. Ernst. 2003. *Snakes of the United States and Canada*. Washington: Smithsonian Institution Press.

Gans, C. 1952. The functional morphology of the egg-eating adaptations in the snake genus *Dasypeltis*. *Zoologica, New York* 37:209–244.

Greene, H. W. 1997. *Snakes: The Evolution of Mystery in Nature*. Berkeley: University of California Press.

Kley, N. J. 2001. Prey transport mechanisms in blindsnakes and the evolution of unilateral feeding systems in snakes. *American Zoologist* 41:1321–1337.

Rossman, D. A., and P. A. Myer. 1990. Behavioral and morphological adaptations for snail extraction in the North American brown snake (genus *Storeria*). *Journal of Herpetology* 24: 434–438.

Sazima, I. 1989. Feeding behavior of the snail-eating snake, *Dipsas indica*. *Journal of Herpetology* 23:464–468.

Why and How Do Snakes Shed Their Skin?

Landmann, L. 1986. Epidermis and dermis. In *Biology of the Integument*. Vol. 2, *Vertebrates*, edited by J. Bereiter-Hahn et al., 150–187. New York: Springer Verlag.

Maderson, P. F. A. 1965. Histological changes in the epidermis of snakes during the sloughing cycle. *Journal of Zoology* 146:98–113.

Zug, G. R., L. J. Vitt, and J. P. Caldwell. 2001. *Herpetology: An Introductory Biology of Amphibians and Reptiles*. 2nd Ed. San Diego: Academic Press.

Why Are Snakes Striped, Banded, or Blotched?

Bechtel, H. B., and E. Bechtel. 1985. Genetics of color mutations in the snake, *Elaphe obsoleta*. *Journal of Heredity* 76:7–11.

———. 1989. Color mutations in the corn snake (*Elaphe guttata guttata*): Review and additional breeding data. *Journal of Heredity* 80:272–276.

Campbell, J. A., and W. W. Lamar. 1989. *The Venomous Reptiles of Latin America*. Ithaca, New York: Comstock Publishing Associates, Cornell University Press.

Cooper, W. E., Jr., and N. Greenberg. 1992. Reptilian coloration and behavior. In *Biology of the Reptilia*, Vol. 18, edited by C. Gans and D. Crews, 298–422. Chicago: University of Chicago Press.

Dyrkacz, S. 1981. Recent instances of albinism in North American amphibians and reptiles. *Society for the Study of Amphibians and Reptiles Herpetological Circular* 11:1–31.

Greene, H. W. 1988. Antipredator mechanisms in reptiles. In *Biology of the Reptilia*, Vol. 16, edited by C. Gans and R. B. Huey, 1–152. New York: Alan R. Liss.

Greene, H. W., and R. W. McDiarmid. 1981. Coral snake mimicry: Does it occur? *Science* 213:1207–1212.

Hedges, S. B., C. A. Hass, and T. K. Maugel. 1989. Physiological color changes in snakes. *Journal of Herpetology* 23:450–455.

Pough, F. H. 1988. Mimicry and related phenomena. In *Biology of the Reptilia*, Vol. 16, edited by C. Gans and R. B. Huey, 153–234. New York: Alan R. Liss.

Rossoti, H. 1985. *Colour*. Princeton, New Jersey: Princeton University Press.

Roze, J. A. 1996. *Coral Snakes of the Americas: Biology, Identification, and Venoms*. Malabar, Florida: Krieger.

Taylor, J. D., and M. E. Hadley. 1970. Chromatophores and color change in the lizard *Anolis carolinensis*. *Z. Zellforsch*. 104:282–294.

How Do Snakes Grow?

Andrews, R. M. 1982. Patterns of growth in reptiles. In *Biology of the Reptilia*, Vol. 13, edited by C. Gans and F. H. Pough, 273–320. London: Academic Press.

Parker, W. S., and M. V. Plummer. 1987. Population ecology. In *Snakes: Ecology and Evolutionary Biology*, edited by R. A. Seigel, J. T. Collins, and S. S. Novak, 253–301. New York: Macmillan.

Shine, R. 1991. *Australian Snakes: A Natural History*. Ithaca, New York: Cornell University Press.

How Long Do Snakes Live?

Conant, R. 1993. The oldest snake. *Bulletin of the Chicago Herpetological Society* 28:77–78.

Fitch, H. S. 1999. *A Kansas Snake Community: Composition and Changes over 50 Years*. Malabar, Florida: Krieger.

Snider, A. T., and J. K. Bowler. 1992. Longevity of reptiles and amphibians in North American Collections, 2nd Ed. *Society for the Study of Amphibians and Reptiles Herpetological Circular* 21:1–40.

What Does It Mean to Be Cold-Blooded?

Harlow, P., and G. Grigg. 1984. Shivering thermogenesis in a brooding diamond python, *Python spilotes spilotes*. *Copeia* 1984:959–965.

Lillywhite, H. B. 1987. Temperature, energetics, and physiological ecology. In *Snakes: Ecology and Evolutionary Biology*, edited by R. A. Seigel, J. T. Collins, and S. S. Novak, 422–477. New York: Macmillan.

Peterson, C. R., A. R. Gibson, and M. E. Dorcas. 1993. Snake thermal ecology: The causes and consequences of body-temperature variation. In *Snakes: Ecology and Behavior*, edited by R. A. Seigel and J. T. Collins, 241–314. New York: Macmillan.

Shine, R. 1988. Parental care in reptiles. In *Biology of the Reptilia*, Vol. 16, edited by C. Gans and R. B. Huey, 275–329. New York: Alan R. Liss.

Van Mierop, L. H. S., and S. M. Barnard. 1978. Further observations on thermoregulation in the brooding female *Python molurus bivittatus* (*Serpentes: Boidae*). *Copeia* 1978:615–621.

How Do Snakes Spend the Winter?

Brown, W. S. 1992. Emergence, ingress, and seasonal captures at dens of northern timber rattlesnakes, *Crotalus horridus*. In *Biology of Pitvipers*, edited by J. A. Campbell and E. D. Brodie, Jr., 251–258. Tyler, Texas: Selva.

Gregory, P. T. 1982. Reptilian hibernation. In *Biology of the Reptilia*, Vol. 13, edited by C. Gans and F. H. Pough, 53–154. London: Academic Press.

Lillywhite, H. B. 1987. Temperature, energetics, and physiological ecology. In *Snakes: Ecology and Evolutionary Biology*, edited by R. A. Seigel, J. T. Collins, and S. S. Novak, 422–477. New York: Macmillan.

Peterson, C. R., A. R. Gibson, and M. E. Dorcas. 1993. Snake thermal ecology: The causes and consequences of body-temperature variation. In *Snakes: Ecology and Evolutionary Biology*, edited by R. A. Seigel, J. T. Collins, and S. S. Novak, 241–314. New York: Macmillan.

How Do Snakes Reproduce?

Butler, J. A., T. W. Hull, and R. Franz. 1995. Neonate aggregations and maternal attendance of young in the eastern diamondback rattlesnake. *Copeia* 1995:196–198.

Carpenter, C. C. 1977. Variation and evolution of stereotyped behaviors in reptiles. Part I. A survey of stereotyped reptilian behavior patterns. In *Biology of the Reptilia*, Vol. 7, edited by C. Gans and D. W. Tinkle, 335–403. London: Academic Press.

Crews, D., and W. R. Garstka. 1982. The ecological physiology of a garter snake. *Scientific American* 1982:159–168.

Crews, D., and M. C. Moore. 1986. Evolution of mechanisms controlling mating behavior. *Science* 231:121–125.

Duvall, D., G. W. Schuett, and S. J. Arnold. 1993. Ecology and evolution of snake mating systems. In *Snakes: Ecology and Evolutionary Biology*, edited by R. A. Seigel, J. T. Collins, and S. S. Novak, 165–200. New York: Macmillan.

Ernst, C. H., and E. M. Ernst. 2003. *Snakes of the United States and Canada*. Washington: Smithsonian Institution Press.

Gillingham, J. C. 1987. Social behavior. In *Snakes: Ecology and Evolutionary Biology*, edited by R. A. Seigel, J. T. Collins, and S. S. Novak, 184–209. New York: Macmillan.

Greene, H. W. 1997. *Snakes: The Evolution of Mystery in Nature*. Berkeley: University of
California Press.

Seigel, R. A., and N. B. Ford. 1987. Reproductive ecology. In *Snakes: Ecology and Evolutionary
Biology*, edited by R. A. Seigel, J. T. Collins, and S. S. Novak, 210–252. New York: Macmillan.

Shine, R. 1988. Parental care in reptiles. In *Biology of the Reptilia*, Vol. 16, edited by C. Gans
and R. B. Huey, 275–329. New York: Alan R. Liss.

———. 1997. The influence of nest temperatures and maternal brooding on hatchling
phenotypes in water pythons. *Ecology* 78:1713–1721.

Do Snakes Have Enemies?

Bauchot, R., ed. 1994. *Snakes: A Natural History*. New York: Sterling.

Dodd, C. K., Jr. 1993. Strategies for snake conservation. In *Snakes: Ecology and Behavior*, edited
by R. A. Seigel and J. T. Collins. New York: McGraw-Hill.

Ernst, C. H., and E. M. Ernst. 2003. *Snakes of the United States and Canada*. Washington:
Smithsonian Institution Press.

Fitzgerald, L. A., and C. W. Painter. 2000. Rattlesnake commercialization: long-term trends,
issues, and implications for conservation. *Wildlife Society Bulletin* 28:235–253.

Klauber, L. M. 1972. *Rattlesnakes: Their Habits, Life Histories, and Influence on Mankind*, 2nd Ed.
Berkeley: University of California Press.

What Sounds Do Snakes Make?

Cook, P. M., M. P. Rowe, and R. W. Van Devender. 1994. Allometric scaling and interspecific
differences in the rattling sound of rattlesnakes. *Herpetologica* 50:358–368.

Fenton, M. B., and L. E. Licht. 1990. Why rattle snake? *Journal of Herpetology* 24:274–279.

Klauber, L. M. 1972. *Rattlesnakes: Their Habits, Life Histories, and Influence on Mankind*, 2nd Ed.
Berkeley: University of California Press.

Young, B. A. 1997. A review of sound production and hearing in snakes, with a discussion of
intraspecific acoustic communication in snakes. *Journal of the Pennsylvania Academy of
Science*. 71:39–46.

Young, B. A., and I. P. Brown. 1993. On the acoustic profile of the rattlesnake rattle. *Amphibia-
Reptilia* 14:373–380.

Zimmerman, A. A., and C. H. Pope. 1948. Development and growth of the rattle of
rattlesnakes. *Fieldiana, Zoology* 32:355–413.

How Do Snakes Defend Themselves?

Engelmann, W., and F. J. Obst. 1981. *Snakes: Biology, Behavior, and Relationships to Man*. New
York: Exeter Books.

Ernst, C. H., and E. M. Ernst. 2003. *Snakes of the United States and Canada*. Washington:
Smithsonian Institution Press.

Greene, H. W. 1988. Antipredator mechanisms in reptiles. In *Biology of the Reptilia*, Vol. 16,
edited by C. Gans and R. B. Huey, 1–152. New York: Alan R. Liss.

———. 1997. *Snakes: The Evolution of Mystery in Nature*. Berkeley: University of California Press.

Pope, C. H. 1958. Fatal bite of captive African rear-fanged snake (*Dispholidus*). *Copeia*
1958:280–282.

Do Snakes Get Sick?

Frye, F. L. 1991. *Reptile Care: An Atlas of Diseases and Treatments*, 2 vols. Neptune, New Jersey:
T.F.H. Publishing.

Murphy, J. B., K. Adler, and J. T. Collins. 1994. *Captive Management and Conservation of
Amphibians and Reptiles*. Oxford, Ohio: Society for the Study of Amphibians and Reptiles.

Messonnier, S. P. 1996. *Common Reptile Diseases and Treatment*. Cambridge, Massachusetts: Blackwell Science.

Reichenbach-Klinke, H., and E. Elkan. 1965. *The Principal Diseases of Lower Vertebrates*. Book III, *Diseases of Reptiles*. London: Academic Press.

Ross, R. A., and G. Marzec. 1984. *The Bacterial Diseases of Reptiles. Their Epidemiology, Control, Diagnosis, and Treatment*. Stanford, California: Institute for Herpetological Research.

When Did the First Snake Appear?

Greene, H. W. 1997. *Snakes: The Evolution of Mystery in Nature*. Berkeley: University of California Press.

Holman, J. A. 2000. *Fossil Snakes of North America: Origin, Evolution, Distribution, Paleoecology*. Bloomington, Indiana: Indiana University Press.

Rage, J. C. 1984. *Serpentes. Handbuch der Paläoherpetologie* II. Stuttgart: Gustav Fischer.

Zug, G. R., L. J. Vitt, and J. P. Caldwell. 2001. *Herpetology: An Introductory Biology of Amphibians and Reptiles*. 2nd Ed. San Diego: Academic Press.

How Many Kinds of Snakes Exist Today?

Cadle, J. E. 1987. Geographic distribution: Problems in phylogeny and zoogeography. In *Snakes: Ecology and Evolutionary Biology*, edited by R. A. Seigel, J. T. Collins, and S. S. Novak, 77–105. New York: Macmillan.

Cundall, D., V. Wallach, and D. A. Rossman. 1994. The systematic relationships of the snake genus *Anomochilus*. *Zoological Journal of the Linnaean Society* 109:275–299.

McDiarmid, R. W., J. A. Campbell, and T. A. Touré. 1999. *Snake Species of the World: A Taxonomic and Geographic Reference. Volume 1*. Washington, D.C.: The Herpetologists' League.

McDowell, S. B. 1987. Systematics. In *Snakes: Ecology and Evolutionary Biology*, edited by R. A. Seigel, J. T. Collins, and S. S. Novak, 3–50. New York: Macmillan.

Zug, G. R., L. J. Vitt, and J. P. Caldwell. 2001. *Herpetology: An Introductory Biology of Amphibians and Reptiles*. 2nd Ed. San Diego: Academic Press.

FOLKTALES

Why So Many Falsehoods?

Bauchot, R., ed. 1994. *Snakes: A Natural History*. New York: Sterling.

Minton, S. A., Jr., and M. R. Minton. 1973. *Giant Reptiles*. New York: Charles Scribner's Sons.

Morris, R., and D. Morris. 1965. *Men and Snakes*. New York: McGraw-Hill.

Shine, R. 1991. *Australian Snakes: A Natural History*. Ithaca, New York: Cornell University Press.

Wilson, E. O. 1984. *Biophilia*. Cambridge, Massachusetts: Harvard University Press.

Which Tales Are True?

Bauchot, R., ed. 1994. *Snakes: A Natural History*. New York: Sterling.

Engelmann, W., and F. J. Obst. 1981. *Snakes: Biology, Behavior, and Relationships to Man*. New York: Exeter Books.

Kitchens, C. S., S. Hunter, and L. H. S. Van Mierop. 1987. Severe myonecrosis in a fatal case of envenomation by the canebrake rattlesnake (*Crotalus horridus atricaudatus*). *Toxicon* 25:455–458.

Klauber, L. M. 1972. *Rattlesnakes: Their Habits, Life Histories, and Influence on Mankind*, 2nd Ed. Berkeley: University of California Press.

Morris, R., and D. Morris. 1965. *Men and Snakes*. New York: McGraw-Hill.

Oliver, J. A. 1958. *Snakes in Fact and Fiction*. New York: Macmillan.

Pope, C. H. 1958. *Snakes Alive and How They Live*. New York: Viking Press.

Schmidt, K. P. 1929. The truth about snake stories. Chicago: Field Museum of the Natural History. Pamphlet.

Can Snakes Be Charmed?

Engelmann, W., and F. J. Obst. 1981. *Snakes: Biology, Behavior, and Relationships to Man*. New York: Exeter Books.

Morris, R., and D. Morris. 1965. *Men and Snakes*. New York: McGraw-Hill.

Oliver, J. A. 1958. *Snakes in Fact and Fiction*. New York: Macmillan.

Pope, C. H. 1958. *Snakes Alive and How They Live*. New York: Viking Press.

GIANT SNAKES: BIG AND BIGGEST

Anacondas

Gilmore, R. M., and J. C. Murphy. 1993. On large anacondas, *Eunectes murinus* (Serpentes: Boidae), with special reference to the Dunn-Lamon record. *Bulletin of the Chicago Herpetological Society* 28:185–188.

Oliver, J. A. 1958. *Snakes in Fact and Fiction*. New York: Macmillan.

Murphy, J. C., and R. W. Henderson. 1997. *Tales of Giant Snakes: A Natural History of Anacondas and Pythons*. Malabar, Florida: Kreiger.

Pope, C. H. 1961. *The Giant Snakes: The Natural History of the Boa Constrictor, the Anaconda, and the Largest Pythons*. New York: Knopf.

Strimple, P. D. 1993. Overview of the natural history of the green anaconda (*Eunectes murinus*). *Herpetological Natural History* 1:25–53.

Reticulated Python

Minton, S. A., Jr., and M. R. Minton. 1973. *Giant Reptiles*. New York: Charles Scribner's Sons.

Murphy, J. C., and R. W. Henderson. 1997. *Tales of Giant Snakes: A Natural History of Anacondas and Pythons*. Malabar, Florida: Kreiger.

Pope, C. H. 1961. *The Giant Snakes: The Natural History of the Boa Constrictor, the Anaconda, and the Largest Pythons*. New York: Knopf.

Smith, M. A. 1943. *The Fauna of British India, Ceylon and Burma, including the Whole of the Indo-China Sub-region. Reptilia and Amphibia*. Vol. III, *Serpentes*. London: Taylor & Francis.

Vasse, Y. 1994. An extraordinary attack. In *Snakes: A Natural History*, edited by R. Bauchot, 110. New York: Sterling.

African Rock Pythons

Branch, W. 1988. *Bill Branch's Field Guide to the Snakes and Other Reptiles of Southern Africa*. Sanibel Island, Florida: Ralph Curtis Books.

Duncan, J. 1847. *Travels in Western Africa in 1845 and 1846. Comprising a Journey from Whydah, through the Kingdom of Dahomey to Adofolldia in the Interior*. London: Richard Bentley

FitzSimons, V. F. M. 1962. *Snakes of Southern Africa*. Capetown, South Africa: Parnell & Sons.

———. 1974. *A Field Guide to the Snakes of Southern Africa*. London: Collins.

Marais, J. 1992. *A Complete Guide to the Snakes of Southern Africa*. Malabar, Florida: Kreiger.

Pitman, C. R. S. 1974. *A Guide to the Snakes of Uganda*, rev. ed. Codicote, England: Weldon & Wesley.

Pope, C. H. 1961. *The Giant Snakes: The Natural History of the Boa Constrictor, the Anaconda, and the Largest Pythons*. New York: Knopf.

Spawls, S., K. Howell, R. Drewes, and J. Ashe. 2002. *A Field Guide to the Reptiles of East Africa.* San Diego: Academic Press.

Indian Python

Daniel, J. C. 1983. *The Book of Indian Reptiles.* Bombay: Bombay Natural History Society.

Pope, C. H. 1961. *The Giant Snakes: The Natural History of the Boa Constrictor, the Anaconda, and the Largest Pythons.* New York: Knopf.

Smith, M. A. 1943. *The Fauna of British India, Ceylon and Burma, including the Whole of the Indo-China Sub-region. Reptilia and Amphibia.* Vol. III, *Serpentes.* London: Taylor & Francis.

Whitaker, R. 1978. *Common Indian Snakes: A Field Guide.* Delhi: Macmillan Company India.

———. 1993. Population status of the Indian python (*Python molurus*) on the Indian subcontinent. *Herpetological Natural History* 1:87–89.

Australian Scrub Python

Cogger, H. G. 1992. *Reptiles and Amphibians of Australia.* Ithaca, New York: Cornell University Press.

Gow, G. F. 1976. *Snakes of Australia.* London: Angus & Robertson.

Kinghorn, J. R. 1956. *The Snakes of Australia.* London: Angus & Robertson.

Shine, R. 1991. *Australian Snakes: A Natural History.* Ithaca, New York: Cornell University Press.

Boa Constrictor

Boos, H. E. A. 1992. A note on the 18.5 ft boa constrictor from Trinidad. *British Herpetological Society Bulletin* 40:15–17.

Ernst, C. H., and E. M. Ernst. 2003. *Snakes of the United States and Canada.* Washington: Smithsonian Institution Press.

Greene, H. W. 1983. *Boa constrictor* (boa, béquer, boa constrictor). In *Costa Rican Natural History,* edited by D. H. Janzen, 380–382. Chicago: University of Chicago Press.

Pope, C. H. 1961. *The Giant Snakes: The Natural History of the Boa Constrictor, the Anaconda, and the Largest Pythons.* New York: Knopf.

Asian Ratsnakes

Cox, M. J., P. P. van Dijk, J. Nabhitabhata, and K. Thirakhupt. 1998. *A Photographic Guide to Snakes and other Reptiles of Peninsular Malaysia, Singapore and Thailand.* Sanibel Isl., Florida: Ralph Curtis Books.

Smith, M. A. 1943. *The Fauna of British India, Ceylon and Burma, including the Whole of the Indo-China Sub-region. Reptilia and Amphibia.* Vol. III, *Serpentes.* London: Taylor & Francis.

Taylor, E. H. 1965. The serpents of Thailand and adjacent waters. *University of Kansas Science Bulletin* 45:609–1096.

Whitaker, R. 1978. *Common Indian Snakes: A Field Guide.* Delhi: Macmillan Company India.

King Cobra

Aagaard, C. J. 1924. Cobras and king cobras. *Journal of the Natural History Society of Siam* 6:315–316.

Smith, H. C. 1936. A hymadryad's (*Naia bungarus*) nest and eggs. *Journal of the Bombay Natural History Society* 39:186.

Smith, M. A. 1943. *The Fauna of British India, Ceylon and Burma, including the Whole of the Indo-China Sub-region. Reptilia and Amphibia.* Vol. III, *Serpentes.* London: Taylor & Francis.

Taylor, E. H. 1965. The serpents of Thailand and adjacent waters. *University of Kansas Science Bulletin* 45:609–1096.

Whitaker, R. 1978. *Common Indian Snakes: A Field Guide.* Delhi: Macmillan Company India.

Taipans

Gow, G. F. 1976. *Snakes of Australia*. London: Angus & Robertson.

Greer, A. E. 1998. *The Biology and Evolution of Australian Snakes*. Chipping Norton, New South Wales: Surrey Beatty & Sons.

Masci, P., and P. Kendall. 1995. *The Taipan: The World's Most Dangerous Snake*. Kenthurst, New South Wales: Kangaroo Press.

Shine, R. 1991. *Australian Snakes: A Natural History*. Ithaca, New York: Cornell University Press.

Bushmasters

Campbell, J. A., and W. W. Lamar. 1989. *The Venomous Reptiles of Latin America*. Ithaca, New York: Comstock Publishing Associates, Cornell University Press.

Greene, H. W. 1986. Natural history and evolutionary biology. In *Predator–Prey Relationships: Perspectives and Approaches from the Study of Lower Vertebrates*, edited by M. Feder and G. C. Lauder, 99–108. Chicago: University of Chicago Press.

Greene, H. W., and M. A. Santana. 1983. Field studies of hunting behavior by bushmasters. *American Zoologist* 23:897.

Ripa, D. 1994. Reproduction of the Central American bushmaster (*Lachesis muta stenophrys*) and the black-headed bushmaster (*Lachesis muta melanocephala*) for the first time in captivity. *Bulletin of the Chicago Herpetological Society* 29:165–183.

———. 1999. Key to understanding the bushmaster (genus *Lachesis* Daudin, 1803). *Bulletin of the Chicago Herpetological Society* 34:45–92.

Zamuda, K. R., and H. W. Greene. 1997. Phylogeography of the bushmaster (*Lachesis muta*: Viperidae): implications for neotropical biogeography, systematics, and conservation. *Biological Journal of the Linnean Society, London* 62:421–442.

Mambas

Broadley, D. G. 1983. *FitzSimons' Snakes*. Johannesburg, South Africa: Delta Books.

Cansdale, G. S. 1961. *West African Nature Handbooks: West African Snakes*. Burn Mill, England: Longman Group.

FitzSimons, V. F. M. 1962. *Snakes of Southern Africa*. Capetown, South Africa: Parnell & Sons.

———. 1974. *A Field Guide to the Snakes of Southern Africa*. London: Collins.

Marais, J. 1992. *A Complete Guide to the Snakes of Southern Africa*. Malabar, Florida: Krieger.

Phelps, T. 1981. *Poisonous Snakes*. Pool, England: Blandford Press.

Pitman, C. R. S. 1974. *A Guide to the Snakes of Uganda*, rev. ed. Codicote, England: Weldon & Wesley.

Spawls, S., K. Howell, R. Drewes, and J. Ashe. 2002. *A Field Guide to the Reptiles of East Africa*. San Diego: Academic Press.

Other Giants and Near-Giants

Boundy, J. 1995. Maximum lengths of North American snakes. *Bulletin of the Chicago Herpetological Society* 30:109–122.

Broadley, D. G. 1983. *FitzSimons' Snakes*. Johannesburg, South Africa: Delta Books.

Campbell, J. A., and W. W. Lamar. 1989. *The Venomous Reptiles of Latin America*. Ithaca, New York: Comstock Publishing Associates, Cornell University Press.

Elliot, W. B. 1978. Chemistry and immunology of reptilian venoms. In *Biology of the Reptilia*, Vol. 8, edited by C. Gans and K. A. Gans, 163–436. London: Academic Press.

Ernst, C. H., and E. M. Ernst. 2003. *Snakes of the United States and Canada*. Washington: Smithsonian Institution Press.

Fritts, T. H. 1988. The brown tree snake, *Boiga irregularis*, a threat to Pacific islands. *U.S. Fish and Wildlife Service Biological Report* 88(31):1–36.

Heatwole, H. 1999. *Sea Snakes*. Malabar, Florida: Krieger.

Moulis, R. 1976. Autecology of the eastern indigo snake, *Drymarchon corais couperi*. *HERP* (New York Herpetological Society) 12:14–23.

Pitman, C. R. S. 1974. *A Guide to the Snakes of Uganda*, rev. ed. Codicote, England: Weldon & Wesley.

Shine, R. 1991. *Australian Snakes: A Natural History*. Ithaca, New York: Cornell University Press.

Spawls, S., K. Howell, R. Drewes, and J. Ashe. 2002. *A Field Guide to the Reptiles of East Africa*. San Diego: Academic Press.

Whitaker, R., and Whitaker, Z. 1998. *Crocodile Fever: Wildlife Adventures in New Guinea*. Chennai: Orient Longman.

SNAKEBITE

Venomous or Poisonous: What Is the Difference?

Edstrom, A. 1992. *Venomous and Poisonous Animals*. Malabar, Florida: Krieger.

Elliot, W. B. 1978. Chemistry and immunology of reptilian venoms. In *Biology of the Reptilia*, Vol. 8, edited by C. Gans and K. A. Gans, 163–436. London: Academic Press.

Ernst, C. H. 1992. *Venomous Reptiles of North America*. Washington: Smithsonian Institution Press.

Greene, H. W. 1997. *Snakes: The Evolution of Mystery in Nature*. Berkeley: University of California Press.

Hill, R. E., and S. P. MacKessy. 1997. Venom yields from several species of colubrid snakes and differential effects of kelamine. *Toxicon* 35:671–678.

Mebs, D. 1978. Pharmacology of reptilian venoms. In *Biology of the Reptilia*, Vol. 8, edited by C. Gans and K. A. Gans, 437–560. London: Academic Press.

Minton, S. A., Jr. 1974. *Venom Diseases*. Springfield, Illinois: Charles C. Thomas.

Minton, S. A., Jr., and M. R. Minton. 1980. *Venomous Reptiles*, rev. ed. New York: Charles Scribner's Sons.

Russell, F. E. 1983. *Snake Venom Poisoning*. Great Neck, New York: Scholium International.

Tu, A. T. 1977. *Venoms: Chemistry and Molecular Biology*. New York: John Wiley & Sons.

———, ed. 1991. *Reptile Venoms and Toxins*. Vol. 5 of *Handbook of Natural Toxins*. New York: Marcel Dekker.

How Many Snakes Are Venomous?

Campbell, J. A., and W. W. Lamar. 1989. *The Venomous Reptiles of Latin America*. Ithaca, New York: Comstock Publishing Associates, Cornell University Press.

Ernst, C. H., and E. M. Ernst. 2003. *Snakes of the United States and Canada*. Washington: Smithsonian Institution Press.

McKinistry, D. M. 1978. Evidence of toxic saliva in some colubrid snakes of the United States. *Toxicon* 16:523–534.

Minton, S. A., Jr. 1990. Neurotoxic snake envenoming. *Seminars in Neurobiology* 10:52–61.

———. 1990. Venomous bites by nonvenomous snakes: A bibliography of colubrid envenomation. *Journal of Wildlife Medicine* 1:119–127.

Shine, R. 1991. *Australian Snakes: A Natural History*. Ithaca, New York: Cornell University Press.

Which Snakes Have the Most Potent Venom?

Ernst, C. H. 1992. *Venomous Reptiles of North America*. Washington: Smithsonian Institution Press.

Heatwole, H. 1999. *Sea Snakes*. Malabar, Florida: Krieger.

Klauber, L. M. 1972. *Rattlesnakes: Their Habits, Life Histories, and Influence on Mankind,* 2nd Ed. Berkeley: University of California Press.

Mebs, D. 1978. Pharmacology of reptilian venoms. In *Biology of the Reptilia,* Vol. 8, edited by C. Gans and K. A. Gans, 437–560. London: Academic Press.

Minton, S. A., Jr. 1989. The most poisonous snakes. League of Florida Herpetological Societies. Unnumbered.

Minton, S. A., Jr., and M. R. Minton. 1980. *Venomous Reptiles,* rev. ed. New York: Charles Scribner's Sons.

Russell, F. E. 1983. *Snake Venom Poisoning.* Great Neck, New York: Scholium International.

Tu, A. T., ed. 1991. *Reptile Venoms and Toxins.* Vol. 5 of *Handbook of Natural Toxins.* New York: Marcel Dekker.

How Do Fangs Work? and How is Venom Injected?

Bogert, C. M. 1943. Dentitional phenomena in cobras and other elapids with notes on adaptive modifications of fangs. *Bulletin of the American Museum of Natural History* 81:285–360.

Carr, A. 1963. *The Reptiles.* New York: Time-Life Books.

Ernst, C. H. 1992. *Venomous Reptiles of North America.* Washington: Smithsonian Institution Press.

Greene, H. W. 1997. *Snakes: The Evolution of Mystery in Nature.* Berkeley: University of California Press.

Kardong, K. V. 1974. Kinesis of the jaw apparatus during the strike in the cottonmouth snake, *Agkistrodon piscivorus. Forma Functio* 7:327–354.

———. 1982. The evolution of the venom apparatus in snakes from colubrids to viperids and elapids. *Memoir du Instituto Butantan* 46:105–118.

Kardong, K. V., and P. A. Lavin-Murcio. 1993. Venom delivery of snakes as high-pressure and low-pressure systems. *Copeia* 1993:644–650.

Klauber, L. M. 1972. *Rattlesnakes: Their Habits, Life Histories, and Influence on Mankind,* 2nd Ed. Berkeley: University of California Press.

Schaefer, N. 1976. The mechanism of venom transfer from the venom duct to the fang in snakes. *Herpetologica* 32:71–76.

Wüster, W., and R. S. Thorpe. 1992. Dentitional phenomena in cobras revisited: Spitting and fang structure in the Asiatic species of *Naja* (Serpentes: Elapidae). *Herpetologica* 48:424–434.

How Dangerous Is a Snakebite?

Campbell, J. A., and W. W. Lamar. 1989. *The Venomous Reptiles of Latin America.* Ithaca, New York: Comstock Publishing Associates, Cornell University Press.

Dart, R., J. T. McNally, et al. 1992. The sequelae of pitviper envenomation in the United States. In *Biology of Pitvipers,* edited by J. A. Campbell and E. D. Brodie, Jr., 395–404. Tyler, Texas: Selva.

de Silva, A., and L. Ranasinghe. 1983. Epidemiology of snake-bite in Sri Lanka: a review. *Ceylon Medical Journal* 28:144–154.

Ernst, C. H. 1992. *Venomous Reptiles of North America.* Washington: Smithsonian Institution Press.

Hardy, D. L. 1992. A review of first aid measures for pitviper bite in North America with an appraisal of Extractor™ Suction and Stun Gun electroshock. In *Biology of Pitvipers,* edited by J. A. Campbell and E. D. Brodie, Jr., 405–414. Tyler, Texas: Selva.

Hardy, D. L., Sr. 2000. Effectiveness of the "Extractor" for pit viper bites questioned. *Notes from NOAH* 28:1–3.

Klauber, L. M. 1972. *Rattlesnakes: Their Habits, Life Histories, and Influence on Mankind,* 2nd Ed. Berkeley: University of California Press.

Mebs, D. 1978. Pharmacology of reptilian venoms. In *Biology of the Reptilia,* Vol. 8, edited by C. Gans and K. A. Gans, 437–560. London: Academic Press.

Minton, S. A., Jr. 1974. *Venom Diseases*. Springfield, Illinois: Charles C. Thomas.

———. 1990. Neurotoxic snake envenoming. *Seminars in Neurobiology* 10:52–61.

———. 1990. Venomous bites by nonvenomous snakes: A bibliography of colubrid envenomation. *Journal of Wildlife Medicine* 1:119–127.

Minton, S. A., Jr., and M. R. Minton. 1980 *Venomous Reptiles*, rev. ed. New York: Charles Scribner's Sons.

Parrish, H. M. 1957. Mortality from snakebites, United States, 1950–1954. *Public Health Reports* 72:1027–1030.

———. 1966. Incidence of treated snakebites in the United States. *Public Health Reports* 81:269–276.

Pope, C. H. 1958. Fatal bite of captive African rear-fanged snake (*Dispholidus*). *Copeia* 1958: 280–282.

Russell, F. E. 1983. *Snake Venom Poisoning*. Great Neck, New York: Scholium International.

Shine, R. 1991. *Australian Snakes: A Natural History*. Ithaca, New York: Cornell University Press.

Swaroop, S., and B. Grab. 1954. Snakebite mortality in the world. *Bulletin of the World Health Organization* 10:35–76.

Tu, A. T. 1977. *Venoms: Chemistry and Molecular Biology*. New York: John Wiley & Sons.

———, ed. 1991. *Reptile Venoms and Toxins*. Vol. 5 of *Handbook of Natural Toxins*. New York: Marcel Dekker.

Warrell, D. A. 1997. Geographical and intraspecies [sic] variation in the clinical manifestations of envenoming by snakes. In *Venomous Snakes: Ecology, Evolution and Snakebite*, edited by R. S. Thorpe, W. Wüster, and A. Malhotra, 189–203. Oxford: Clarendon Press.

What Is Antivenom?

Greene, H. W. 1997. *Snakes: The Evolution of Mystery in Nature*. Berkeley: University of California Press.

Latifi, M. 1991. *The Snakes of Iran*. Oxford, Ohio: Society for the Study of Amphibians and Reptiles.

Tu, A. T., ed. 1991. *Reptile Venoms and Toxins*. Vol. 5 of *Handbook of Natural Toxins*. New York: Marcel Dekker.

Can Venom Be Used as Medicine?

Roze, J. A. 1996. *Coral Snakes of the Americas: Biology, Identification, and Venoms*. Malabar, Florida: Krieger.

Russell, F. E. 1983. *Snake Venom Poisoning*. Great Neck, New York: Scholium International.

SNAKES AND US

Why Are Snakes Important to Us?

Dodd, C. K., Jr. 1987. Status, conservation, and management. In *Snakes: Ecology and Evolutionary Biology*, edited by R. A. Seigel, J. T. Collins, and S. S. Novak, 478–513. New York: Macmillan.

———. 1993. Strategies for snake conservation. In *Snakes: Ecology and Behavior*, edited by R. A. Seigel and J. T. Collins, 363–393. New York: McGraw-Hill.

Stoker, K. F., ed. 1990. *Medical Use of Snake Venom Proteins*. Boca Raton, Florida: CRC Press.

Why Do I Have Snakes in My House?

Rodda, G. H., Y. Sawai, D. Chiszar, and H. Tanaka. 1999. *Problem Snake Management: The Habu and the Brown Treesnake*. Ithaca, New York: Cornell University Press.

Tanaka, H., Y. Sawai, et al. 1985. Improvement of control methods of habu, *Trimeresurus flavoviridis,* the venomous snakes on the Amani Islands . . . from FY 1977 to 1979. *Snake* 17:96–111.

Should I Keep a Snake as a Pet?

Frye, F. L. 1991. *Reptile Care: An Atlas of Diseases and Treatments,* 2 vols. Neptune, New Jersey: T.F.H. Publishing.

Murphy, J. B., K. Adler, and J. T. Collins. 1994. *Captive Management and Conservation of Amphibians and Reptiles.* Oxford, Ohio: Society for the Study of Amphibians and Reptiles.

Trutnau, L. 1986. *Nonvenomous Snakes: A Comprehensive Guide to Care and Breeding of over 100 Species.* New York: Barron's Educational Series.

How Can I Become a Snake Specialist?

Zug, G. R. 1989. Careers in herpetology. Gainesville, Florida: American Society of Ichthyologists and Herpetologists. Pamphlet.

What Remains to Be Learned about Snakes?

Dodd, C. K., Jr. 1987. Status, conservation, and management. In *Snakes: Ecology and Evolutionary Biology,* edited by R. A. Seigel, J. T. Collins, and S. S. Novak, 478–513. New York: Macmillan.

———. 1993. Strategies for snake conservation. In *Snakes: Ecology and Behavior,* edited by R. A. Seigel and J. T. Collins, 363–393. New York: McGraw-Hill.

Greene, H. W. 1997. *Snakes: The Evolution of Mystery in Nature.* Berkeley: University of California Press.

Shine, R. 1991. *Australian Snakes: A Natural History.* Ithaca, New York: Cornell University Press.

Seigel, R. A., and J. T. Collins. 1993. *Snakes: Ecology and Behavior.* New York: McGraw-Hill.

Seigel, R. A., J. T. Collins, and S. S. Novak. 1987. *Snakes: Ecology and Evolutionary Biology.* New York: Macmillan.

APPENDIX 1. CLASSIFICATION OF SNAKES

Cadle, J. E. 1987. Geographic distribution: Problems in phylogeny and zoogeography. In *Snakes: Ecology and Evolutionary Biology,* edited by R. A. Seigel, J. T. Collins, and S. S. Novak. New York: Macmillan.

Crother, B. I. (coord.) 2000. Scientific and Standard English Names of Amphibians and Reptiles of North America North of Mexico, with Comments regarding Confidence in our Understanding. *Herpetological Circular,* No. 29.

David, P., and I. Ineich. 1999. Les serpents venimeux du monde: systématique et répartition. *Dumerilia,* Vol. 3.

Kluge, A. G. 1991. Boine Snake Phylogeny and Research Cycles. *University of Michigan Museum of Zoology Miscellaneous Publications,* No. 178.

———. 1993. *Aspidites* and the phylogeny of pythonine snakes. *Records of the Australian Museum,* Supplement 19.

McDiarmid, R., J. A. Campbell, and T. Touré. 1999. *Snake Species of the World: A Taxonomic and Geographic Reference. Volume 1.* Washington, D.C.: The Herpetologists' League.

McDowell, S. B. 1987. Systematics. In *Snakes: Ecology and Evolutionary Biology,* edited by R. A. Seigel, J. T. Collins, and S. S. Novak, 3–50. New York: Macmillan.

Underwood, G. 1967. *A Contribution to the Classification of Snakes.* London: British Museum (Natural History).

Zug, G. R., L. J. Vitt, and J. P. Caldwell. 2001. *Herpetology: An Introductory Biology of Amphibians and Reptiles.* 2nd Ed. San Diego: Academic Press.

TAXONOMIC INDEX

Common names are listed by the "core" part of the common name. Species are also listed by their scientific names. Page numbers in *italics* refer to photographs or figures.

SUBJECT INDEX

Page numbers in *italics* refer to photographs or figures.

seeing
 infrared vision, 17–18, *17*
 overview, 15–17
senses. *See also specific senses*
 overview, 15–20
sensory capsules, 5
sidewinding, 11–12, *11*
size
 giants/near-giants, *80*, 81–103, *82–83*, *85*,
  87–90, *92–95*, *97–98*, *100–2*
 growth, 34–36
skin
 description/function, 5, 27
 shedding/replacement, 27–29, *28*, *29*, 34
skulls, 5–6, *7*
slidepushing, 12
smelling. *See* odor
snakebite. *See also* venom; venomous snakes
 from decapitated rattlesnake, 74, 124
 "dry" bites, 125
 factors affecting seriousness, 113, 122, *123*
 medical attention, 122, 123, *123*, 124–25
 steps following, 124–25
 venomous snakes/deaths, 122–24
snake charmers, 78–79, *78*
snakes
 commercial exploitation of, 131–32, *131*
 description/traits overview, 1–8
 importance to humans, 129–32
 lizards vs., 1–3, *2*
 studying, 36, 139–40, *139*
snakes in human houses/yards
 controlling, 134–36, *136*
 reasons for, 132–33, *133*, *134*
 repellents/removal, 132, 134, 135–36, *136*
snake specialists, 36, 139
solenoglyphous fangs, 117–18, *117*, 118–19, *119*, 120
sounds made by snakes, 31, 52–54, *53*
species definition, 64
spectacle, 2, 28
stewardship, 129–30, 132
swimming, 10–11, *10*

tasting food, 20
teeth. *See also* fangs
 function, 6, 24
tongue flicking, 19, *19*, 20
tracheal lung, 2, 8

undulatory locomotion, 10–11, 13
uric acid, 7

venom. *See also* antivenom
 delivery systems, *115*, 116–20, *117*, *119*
 description/chemistry, 105–8
 evolution, 22, 105–6, 108
 fangs, 113, 115–20, *115*, *117*, *119*
 medical use, 126–27, 132
 neurotoxic vs. hemotoxic, 106
 newborn/juveniles, 108, 120
 poison vs., 105
 prey capture/handling, 22, 23, 105, *106*, 126
 spitting of, 55, 99, *100*, 120–22, *121*
volume injected, 108, 112, 113
venom glands, 105, 115
venomous snakes
 distribution/location, 110–12, *110*
 interbreeding with nonvenomous snakes,
  76–77, *77*
 overview, 108–9
 pets, 123–24, 138–39
 potency/toxicity, 108, *109*, 112–13, *112*, 114
 snakebite deaths, 122–24
 triangular heads/hoods, *72*, 73–74
venom races, 107–8, *107*
vertebrae, 3, 4, 5
vertebral plexus, 7–8
vertebrate traits, 1
viral diseases, 57, 58
vomeronasal (Jacobson's) organ, 19, 43

water needs (hibernation), 41
water submergence times, 8–9

SMITHSONIAN ANSWER BOOKS

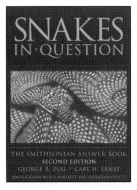

Paperback $24.95
ISBN: 1-58834-114-3
Hardcover $55.00
ISBN: 1-58834-113-5

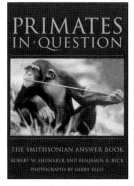

Paperback $27.95
ISBN: 1-58834-176-3
Hardcover $55.00
ISBN: 1-58834-151-8

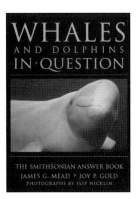

Paperback $24.95
ISBN: 1-56098-980-7
Hardcover $55.00
ISBN: 1-56098-955-6

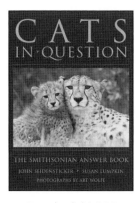

Paperback $24.95
ISBN: 1-58834-126-7
Hardcover $55.00
ISBN: 1-58834-125-9

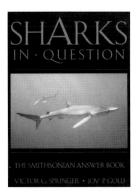

Paperback $24.95
ISBN: 0-87474-877-1

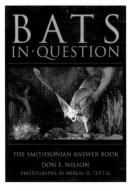

Paperback $24.95
ISBN: 1-56098-739-1

Smithsonian Books